ILEANA TOMA
VALERICA MOŞNEGUŢU
ŞTEFANIA CONSTANTINESCU

ORDINARY DIFFERENTIAL EQUATIONS AND SYSTEMS

AN INTRODUCTION, WITH APPLICATIONS AND EXERCISES

SERIES: MATHEMATICS FOR FUTURE ENGINEERS

VOLUME 3

Copyright © 2012 Ileana Toma
All rights reserved.

ISBN-10: 154031801X
ISBN-13: 978-1540318015

CONTENTS

PREFACE iii

CHAPTER 1 1

FIRST ORDER ORDINARY DIFFERENTIAL EQUATIONS 1

1.1. Preliminary notions. Examples 1

1.2. The forms of the first order ODEs and of their solutions 9

    1.2.1. Forms of the first order ODEs 9

    1.2.2. Forms of the solutions 11

1.3. Types of first order ODEs solvable by quadratures 14

    1.3.1. Equations with separate variables 14

    1.3.2. Equations with separable variables 15

    1.3.3. Homogeneous ODEs of order n 16

    1.3.4. Exact/total differential equations 20

    1.3.5. Integrating factor 25

    1.3.6. First order linear ODEs 32

    1.3.7. Bernoulli's equation 42

    1.3.8. Riccati's equation 45

    1.3.9. Clairaut's equation 47

    1.3.10. Lagrange's equation 51

1.4. The method of successive approximation 56

    1.4.1. Cauchy-Picard's theorem of existence and uniqueness 56

    * 1.4.2. The contraction mapping principle 62

Exercises and problems 72

CHAPTER 2 85

LINEAR ORDINARY DIFFERENTIAL EQUATIONS OF ORDER n 85

2.1. Preliminary concepts. Examples 85

2.2. Linear homogeneous ODEs of order n 89

2.3. Linear non-homogeneous ODEs of nth order 103

2.4. Linear ODEs of order n, with constant coefficients 111

    2.4.1. Linear homogeneous ODEs 111

2.4.2. Differential polynomial 123

2.4.3. Non-homogeneous linear ODEs 126

2.5. Equations reducible to ODEs with constant coefficients 139

Exercises and problems 145

CHAPTER 3 .... **153**

SYSTEMS OF LINEAR ODES WITH CONSTANT COEFFICIENTS 153

3.1. Matrix form of the system 153

3.2. Linear homogeneous ODSs 154

3.3. Linear non-homogeneous systems 166

3.4. The link between ODEs and ODSs 168

Exercises and problems 181

REFERENCES 185

Preface

This book is addressed to all those who, after finishing the high school, wish a practical initiation in the domain of ordinary differential equations and systems.

It has its origins in a course written for the first year students at the Technical University of Constructions of Bucharest. As this is an introduction to ordinary differential equations, we tackled only linear ODEs of high order and only linear ODSs with two or three equations.

To provide useful tools for (future) engineers and for specialists, in general, we established mathematical models for some elementary problems of motion and of statics of constructions, by using ordinary differential equations. To this end, we indicate the standard steps to be followed in constructing a mathematical model of a physical phenomenon.

We tried to make the involved mathematics as attractive as possible, by simplifying the presentation without loosing the mathematical rigor of the results. To increase accessibility and to encourage the reader to get a technical know-how about ordinary differential equations, we provided for each newly introduced notion a series of applications and solved problems; each chapter ends by a section containing exercises and problems, each one of these being accompanied by hints and answers.

The sections marked by asterisks can be omitted, as well as some proofs. We introduced them, though, for the sake of a unitary and logical presentation.

The references contain, along with the books, some links with useful sites, which can be helpful for the reader.

Among the books, we especially mention *Ordinary Differential Equations with applications to mechanics*, written by M.V.Soare, P.P.Teodorescu, Ileana Toma, a volume appeared at Springer in 2006. This book can be consulted and study for the purpose of a practical use of ordinary differential equations as models in mechanics and engineering. The applications are structured following four steps: 1) physical phenomenon, 2) mathematical model, 3) solution and 4) physical interpretation of the solution.

The authors are grateful to prof. Gavriil Păltineanu from the Technical University of Civil Engineering of Bucharest, Department of Mathematics and Computer Science, for a thoroughful analysis of the content of this book.

The Authors

Chapter 1

FIRST ORDER ORDINARY DIFFERENTIAL EQUATIONS

1.1. PRELIMINARY NOTIONS. EXAMPLES

The algebraic equation is a notion familiar with the reader.

A *differential equation* is also an equality allowing a *function* as unknown quantity and containing its *derivatives* too. We distinguish two possibilities:

a) the unknown function depends only on one variable, then we have an ***ordinary differential equation*** (abbrev. **ODE**);

b) the unknown function depends on several variables. In this case, we have a ***partial differential equation*** (abbrev. **PDE**).

In this book, we consider only the case *a)*.

As previously mentioned, ***the general form*** of an ODE is

$$F\left(x, y, y', y'', \ldots, y^{(n)}\right) = 0. \qquad (1.1.1)$$

Definition 1.1. The order of an ODE coincides with the maximum order of differentiation of the unknown function y.

One of the fundamental issues of the differential calculus is to compute the derivative of a given function. The simplest inverse problem belongs to the integral calculus:

PROBLEM. *Given a real function $f = f(x)$, of one real variable, find its primitive.*

If we denote the primitive of *f* by *y*, then the mathematical expression of this problem is:

$$\frac{dy}{dx} = f(x), \qquad (1.1.2)$$

or, equivalently

$$dy = f(x)dx. \qquad (1.1.3)$$

The above-mentioned relations are, in fact, the simplest differential equations and we know how to solve them. Indeed, we know that the most general function *y*, satisfying (1.1.2) or (1.1.3), is

$$y(x) = \int f(x)dx + C. \qquad (1.1.4)$$

An arbitrary primitive of *f* is, therefore, a **solution** of the equation (1.1.2). Introducing it in (1.1.2), we get an identity.

Therefore, in the case of the ODE too, a **solution transforms the equation into an identity**, precisely as in the case of algebraic equations.

In the expression (1.1.4), the sign $\int$ designates one of the **primitives** of *f* and *C* is an **arbitrary constant**. Therefore, the function *y* is not uniquely determined by the equation (1.1.2) or

(1.1.3), so that we can state that these equations admit infinitely many solutions. Each solution can be determined by giving to C various numerical values.

Terminology

- ♣ The solution (1.1.4) of the equation (1.1.2) is called the ***general solution***.
- ♣ Any solution obtained by giving particular values to C in the general solution is called a ***particular solution***.
- ♣ Any other solution is called a ***singular solution***.

After all that, it seems that the differential equations appeared in a pure mathematic frame, as a formal logical consequence of the differential calculus.

Yet, this mathematical domain has its historical roots in Newtonian mechanics. Newton, who, together with Leibniz, is the pioneer of the differential calculus, has modelled a series of physical phenomena by using differential equations, with a remarkable intuition. Thus, the famous second law (of classical mechanics), is briefly stated as:

"The vector sum of the external forces acting upon a system is equal to the mass of the system multiplied by its acceleration."

This law, named after Newton, is expressed in mathematical terms as:

$$m\mathbf{a} = \mathbf{F}, \qquad (1.1.5)$$

which is nothing else than a *system of ordinary differential equations*. Indeed, the acceleration is the second derivative of the displacement with respect to time; this specification belongs to another giant of science, namely to Leonhard Euler.

In order to illustrate this, we shall follow the path suggested by Newton when studying an extremely simple case.

PROBLEM. *Study the motion of a particle (material point) M, acted upon by its own weight along a vertical axis.*

Solution. Firstly, we create the **mathematical model**. Therefore, we must establish

a) the unknown function (or functions), whose knowledge means the knowledge of the phenomenon;

b) the physical law (or laws) governing the phenomenon.

Suppose that Oy is the vertical axis along which the particle falls, with its origin at the earth surface (see the figure below).

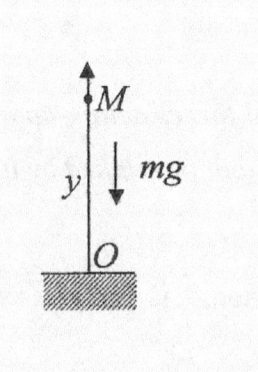

The particle movement is known if its unique coordinate is given – specifically, its position y on the Oy-axis – at every moment t. Therefore, the **unknown function** of the problem is $y = y(t)$, having the physical significance of **displacement** in physics. In problems of motion, **Newton's second law** plays an important part. If we apply it for the unique component of the acceleration, we get

$$ma = -mg, \qquad (1.1.6)$$

where m is the mass particle and g – the acceleration of gravity. The minus comes from the fact that the Oy axis is directed upward and the force of gravity – downward. Taking into account that the acceleration is the second derivative of the displacement with respect to the time t and simplifying by m, it follows that

$$\frac{d^2 y}{dt^2} = -g. \qquad (1.1.7)$$

The equation (1.1.7) represents the **mathematical model** of the considered physical phenomenon. Its mathematical meaning is:

Given the second derivative of the function y, find y.

In this case, this request does not ask for special knowledge. Indeed, taking twice the primitive of both members of the equation (1.1.7), we respectively get

$$\frac{dy}{dt} = -gt + C_1,$$
$$y(t) = -\frac{gt^2}{2} + C_1 t + C_2. \qquad (1.1.8)$$

The last expression is the **general solution of** the equation (1.1.7).

Remark. In this case, the general solution depends on two arbitrary constants, while in case of the equation (1.1.2), it depends only on one.

IMPORTANT!
The general solution of a differential equation always depends on a number of constants equal to the maximum order of derivation of the unknown function.

We shall present later an explanation for this significant fact.

Now, let us specify the meaning of the constants C_1 and C_2 in physics. By putting $t = 0$ in the first expression (1.1.8), we get

$$C_1 = \left.\frac{dy}{dt}\right|_{t=0} = v_0, \qquad (1.1.9)$$

where v_0 is the **initial velocity** of the particle. Similarly, from the second expression (1.1.8) we infer that

$$C_2 = y(t)\big|_{t=0} = y_0, \qquad (1.1.10)$$

which is the **initial position** of the particle.

With these new notations of constants – suggestive, by their physical significance – the general solution of the equation (1.1.7) takes the form

$$y(t) = -\frac{gt^2}{2} + v_0 t + y_0, \qquad (1.1.11)$$

which is familiar to the reader from the high school.

It is clear now which are the supplementary data that must be known in order to find that solution, which corresponds to a well-defined motion:

♣ ***the initial position*** y_0 of the particle and

♣ its ***initial velocity*** v_0.

Therefore, one can say that y satisfies the conditions

$$y(0) = y_0,$$
$$\frac{dy}{dt}(0) = v_0. \qquad (1.1.12)$$

These are also called ***initial conditions*** or ***Cauchy conditions***.

The problem which consists in solving the equation (1.1.7), so that y would satisfy the initial conditions (1.1.12), is called a ***Cauchy problem*** or an ***initial problem***.

IMPORTANT!
In a Cauchy problem, the conditions are established at the same point!

(In the above example, at the point $t = 0$).

However, there are situations in which this type of conditions does not correspond to the physical phenomenon. Let us consider the case of a simply supported bar (see the figure below).

The problem consists in finding the ***deflexion*** (bending) y as a function of x. This time, we shall not present into detail the associated mathematical model. We only specify that it is of the form of an ordinary differential equation

$$\frac{d^2 y}{dx^2} = f(x)\left[1 + \left(\frac{dy}{dx}\right)^2\right]^{\frac{3}{2}}, \qquad (1.1.13)$$

also known as *the Bernoulli-Euler equation*.

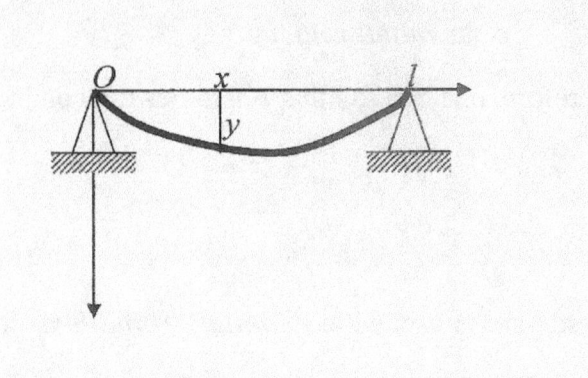

The figure shows that the deflexion must be null at both bar ends 0 and l, i.e.

$$y(0)=0, \qquad y(l)=0. \qquad (1.1.14)$$

The additional conditions (1.1.14) are also called **bilocal** or **two-point conditions**.

The problem which consists in solving the equation (1.1.13) with the conditions (1.1.14) is called a **two-point problem** or a **Picard problem**.

These two types of conditions associated to an ODE are standard and cover an important part of the fundamental mechanical and physical problems.

Following the above-mentioned considerations, we conclude that a systematic study of a physical phenomenon compulsory requires the set-up and use of its differential model.

After solving the ODE (or PDE) model, the interpretation of the solution will allow the effective knowledge, the prevision and, therefore, the control of the studied phenomenon; these are the major targets of science.

1.2. THE FORMS OF THE FIRST ORDER ODEs AND OF THEIR SOLUTIONS

Obviously, an ordinary differential equation works only at the points at which it is defined. For example, the equation

$$y' = \sqrt{1-y^2} \qquad (1.2.1)$$

makes sense only for $|y| \leq 1$.

Given an ODE, one must firstly determine its domain of definition; the domain of definition of an ODE is that of the functions which define it.

1.2.1. FORMS OF THE FIRST ORDER ODEs

A. According to the definition 1.1 and the relation (1.1.1), ***the general form*** of the first order ordinary differential equations is

$$F(x, y, y') = 0, \qquad y' = \frac{dy}{dx}, \qquad (1.2.2)$$

where F is defined – and, usually, continuous – with respect to the independent variable x, as well as with respect to the unknown function y and to its derivative y'.

Its general form is also called ***implicit***, because it includes y' implicitly.

B. If $\dfrac{\partial F}{\partial y'} \neq 0$, then, according to the implicit function theorem (see e.g. [2][3][5][10]), y' can be explicited from (1.2.2) and we obtain the **canonical form** of the first-order ODE:

$$y' = f(x, y), \qquad (1.2.3)$$

also known as its **explicit** form.

C. If $f(x, y) \neq 0$, then (1.2.3) can be also written as

$$\frac{dx}{dy} = \frac{1}{f(x, y)}, \qquad (1.2.4)$$

known as the **inverse form**, which can be used in the neighbourhood of those points $(x, y) \in \Re^2$ at which $f(x, y)$ tends to infinity. Obviously, if f does not tend to infinity, the forms (1.2.3) and (1.2.4) are equivalent.

D. One can also write the ODE (1.2.3) in **differential form**:

$$dy = f(x, y) dx, \qquad (1.2.5)$$

an equivalent to (1.2.3), (1.2.4) form. The more general **differential form**

$$P(x, y) dx + Q(x, y) dy = 0, \qquad (1.2.6)$$

is equivalent too to each of the equations

$$\frac{dy}{dx} = -\frac{P(x, y)}{Q(x, y)}, \qquad \frac{dx}{dy} = -\frac{Q(x, y)}{P(x, y)}. \qquad (1.2.7)$$

WARNING!

At the points (x_0, y_0) at which both P and Q cancel, none of the equations (1.2.6), (1.2.7) *is defined*.

The functions *P* and *Q* are usually continuous on the domain of definition, in the case of the equation (1.2.2).

E. *The symmetric form* of the first order ODE is

$$\frac{dx}{X(x,y)} = \frac{dy}{Y(x,y)}. \qquad (1.2.8)$$

Each of the above-mentioned forms emphasizes several characteristics and possibilities of solving first order equations. The most frequently used in practice are the forms (1.2.2), (1.2.3) and (1.2.6).

1.2.2. FORMS OF THE SOLUTIONS

Definition 1.2. A *solution of the differential equation* (1.2.3) in the real interval $[a,b]$ is a function $y = y(x)$ of class $C^1([a,b])$, which identically satisfies (1.2.3), i.e.

$$y'(x) = f(x, y(x)), \qquad x \in [a,b]. \qquad (1.2.9)$$

If there is a constant *c* such that $f(x,c) = 0$ for any $x \in [a,b]$, we obviously get $y = c$ as a solution of (1.2.3). This is called a ***stationary solution*** and it is particularly important for the qualitative study of the equation.

To solve a first order ODE, one uses, as the case may be, the forms specified in the previous paragraph and, consequently, we shall obtain their solutions in different forms.

The solutions of a first order ODE can be found

a. in *implicit form*: $\Phi(x, y) = 0$;

b. in *explicit form*: $y = y(x)$, $x \in [a, b]$;

c. in *parametric form*: $\begin{cases} x = x(t), \\ y = y(t), \end{cases}$ $t \in [a, b] \subseteq \Re$.

Example. The function

$$y = \sqrt{1 - x^2}, \qquad x \in (-1, 1), \tag{1.2.10}$$

is the *explicit solution* of the equation

$$y' = -\frac{x}{y}. \tag{1.2.11}$$

VERIFICATION. Indeed, on the one hand

$$\frac{dy}{dx} = \frac{d}{dx}\left(\sqrt{1-x^2}\right) = \frac{-2x}{2\sqrt{1-x^2}} = -\frac{x}{\sqrt{1-x^2}}, \tag{1.2.12}$$

and on the other hand,

$$-\frac{x}{y} = -\frac{x}{\sqrt{1-x^2}}. \tag{1.2.13}$$

The expressions (1.2.12) and (1.2.13) coincide.

The solution (1.2.10) can also be expressed in an *implicit form*:

$$\Phi(x, y) \equiv x^2 + y^2 - 1 = 0. \tag{1.2.14}$$

VERIFICATION. Computing the differential of Φ, we get

$$d\Phi(x,y) = d(x^2 + y^2 - 1) = 2x\,dx + 2y\,dy = 0. \tag{1.2.15}$$

From the last equality, we get

$$y' + \frac{x}{y} = 0, \tag{1.2.16}$$

i.e. (1.2.11).

The solution (1.2.10) can be also expressed in *parametric form*:

$$\begin{cases} x = \cos t, \\ y = \sin t, \end{cases} \quad t > 0. \tag{1.2.17}$$

VERIFICATION. We can write the equation (1.2.11) in differential form

$$x\,dx + y\,dy = 0. \tag{1.2.18}$$

We have

$$\begin{aligned} x\,dx &= \cos t \cdot d(\cos t) = \cos t \cdot (-\sin t\,dt) = \\ &= -\sin t \cos t\,dt \\ y\,dy &= \sin t \cdot d(\sin t) = \sin t \cdot (\cos t\,dt) = \\ &= \sin t \cos t\,dt \end{aligned} \quad + \tag{1.2.19}$$

$$x\,dx + y\,dy = (-\sin t \cos t + \sin t \cos t)\,dt,$$

therefore we obtain $x\,dx + y\,dy = 0$, which is precisely (1.2.16).

1.3. TYPES OF FIRST ORDER ODEs SOLVABLE BY QUADRATURES

There are some equations of particular forms, frequently used in applications, that could be solved by certain methods allowing to express the solution in terms of primitives of functions. We say, in these cases, that *the equation is solved by quadratures* (integrations).

In what follows, we shall mention and solve several such types of differential equations.

1.3.1. EQUATIONS WITH SEPARATE VARIABLES

They are of the form

$$X(x)\,dx + Y(y)\,dy = 0, \qquad (1.3.1)$$

where X and Y are continuous functions depending on the variables x and y.

METHOD OF SOLVING

We notice that the function

$$F(x,y) = \int X(x)\,dx + \int Y(y)\,dy, \qquad (1.3.2)$$

gives, by differentiation, the left side of the equation (1.3.1). Indeed,

$$dF(x,y) = \frac{\partial F}{\partial x}dx + \frac{\partial F}{\partial y}dy = X(x)\,dx + Y(y)\,dy. \qquad (1.3.3)$$

Therefore $dF(x,y) = 0$, which leads to $F(x,y) = C$.

Consequently, the general solution of the ODE (1.3.1) is

$$\int X(x)\,dx + \int Y(y)\,dy = C. \qquad (1.3.4)$$

Example. Solve the equation

$$\underbrace{e^{-x}}_{X(x)}dx + \underbrace{\frac{1}{y}}_{Y(y)}dy = 0. \qquad (1.3.5)$$

Solution. It is, obviously, an equation with separate variables. Computing the primitives, one finds

$$\int X(x)\,dx = \int e^{-x}dx = -e^{-x},$$
$$\int Y(y)\,dy = \int \frac{1}{y}dy = \ln|y|, \qquad (1.3.6)$$

thus the general solution of the ODE (1.3.5) is

$$\boxed{-e^{-x} + \ln|y| = C}. \qquad (1.3.7)$$

1.3.2. EQUATIONS WITH SEPARABLE VARIABLES

They are of the form

$$P(x)q(y)\,dx + Q(y)p(x)\,dy = 0, \qquad (1.3.8)$$

where P, p, Q, q are continuous functions with respect to their corresponding arguments.

METHOD OF SOLVING

If $\mu(x, y) \equiv p(x)q(y) \neq 0$ on the domain of definition of the equation, we divide by μ, thus getting

$$\frac{P(x)}{p(x)}dx + \frac{Q(y)}{q(y)}dy = 0, \qquad (1.3.9)$$

which is an equation with separate variables. According to the previous case, the general solution of the equation (1.3.8) is

$$\int \frac{P(x)}{p(x)} dx + \int \frac{Q(y)}{q(y)} dy = C. \qquad (1.3.10)$$

Example. Solve the equation

$$2x\sqrt{y}\, dx + (1-x^2)dy = 0. \qquad (1.3.11)$$

Solution. It is an ODE with separable variables. We divide by $\mu = (1-x^2)\sqrt{y}$ and, after simplification, we obtain

$$\frac{2x}{1-x^2} dx + \frac{1}{\sqrt{y}} dy = 0. \qquad (1.3.12)$$

This is an equation with separate variables, therefore the general solution is given by

$$\int \frac{2x}{1-x^2} dx + \int \frac{1}{\sqrt{y}} dy = C, \qquad (1.3.13)$$

or, by computing the primitives,

$$-\ln|1-x^2| + 2\sqrt{y} = C, \qquad (1.3.14)$$

valid for $1-x^2 \neq 0$, $y > 0$.

1.3.3. HOMOGENEOUS ODEs OF ORDER m

Definition 1.3. A function $f = f(x, y)$, $f : \Re^2 \to \Re$, is said to be *homogeneous of order m* if:

$$f(tx, ty) = t^m f(x, y), \ \forall t \in \Re. \qquad (1.3.15)$$

If the equality occurs only for $t > 0$, f is said to be positively homogeneous.

A first order homogeneous ODE is of the form

$$P(x, y)\,dx + Q(x, y)\,dy = 0, \qquad (1.3.16)$$

where P and Q are both homogeneous, of the same order m.

METHOD OF SOLVING

We make the change

$$y = xz \rightarrow dy = x\,dz + z\,dx. \qquad (1.3.17)$$

Introducing these expressions into the equation, we get

$$P(x, xz)\,dx + Q(x, xz)(x\,dz + z\,dx) = 0. \qquad (1.3.18)$$

But P, Q are homogeneous of order m, hence

$$P(x, xz) = x^m P(1, z), \quad Q(x, xz) = x^m Q(1, z). \qquad (1.3.19)$$

Consequently,

$$x^m \left[\, P(1, z)\,dx + Q(1, z)(x\,dz + z\,dx)\,\right] = 0. \qquad (1.3.20)$$

From this, we have

$$\left[P(1, z) + zQ(1, z)\right]dx + xQ(1, z)\,dz = 0. \qquad (1.3.21)$$

This is an *ODE with separable variables*.

We divide by $x\left(P(1, z) + zQ(1, z)\right)$ and we obtain

$$\frac{dx}{x} + \frac{Q(1, z)}{P(1, z) + zQ(1, z)}\,dz = 0; \qquad (1.3.22)$$

according to the previous considerations, the general solution of the ODE (1.3.21) is

$$\int \frac{dx}{x} + \int \frac{Q(1,z)}{P(1,z)+zQ(1,z)} dz = C. \quad (1.3.23)$$

Using the notation

$$\varphi(z) = \int \frac{Q(1,z)}{P(1,z)+zQ(1,z)} dz, \quad (1.3.24)$$

the general solution of the ODE (1.3.22) can be written as

$$\ln|x| + \varphi(z) = C, \quad (1.3.25)$$

or, passing to exponentials,

$$x = Ce^{-\varphi(z)}. \quad (1.3.26)$$

Turning back to the initial variables, we get the general solution of the homogeneous ODE (1.3.16) in the form

$$x = Ce^{-\varphi\left(\frac{y}{x}\right)}. \quad (1.3.27)$$

Example. Find the general solution of the following homogeneous ODE:

$$(xy + y^2)dx - (2x^2 + xy)dy = 0. \quad (1.3.28)$$

Solution. We have $P(x,y) = xy + y^2$ and $Q(x,y) = -(2x^2 + xy)$.

Obviously, this equation is neither with separate variables, nor with separable variables.

Let us try to check whether it is or not homogeneous, according to the definition 1.3. We have:

$$P(xt,yt) = t^2xy + t^2y^2 = t^2(xy+y^2) = t^2P(x,y),$$
$$Q(xt,yt) = -(2t^2x^2 + t^2xy) = -t^2(2x^2+xy) = \quad (1.3.29)$$
$$= t^2Q(x,y).$$

Therefore the equation is **homogeneous of second order**.

To solve it, we make the change

$$y = xz \rightarrow dy = xdz + zdx. \quad (1.3.30)$$

We successively deduce

$$(x^2z + x^2z^2)dx - (2x^2 + x^2z)(xdz + zdx) = 0,$$
$$x^2\left[(z+z^2)dx - (2+z)(xdz + zdx)\right] = 0,$$
$$\left[(z+z^2) - 2z - z^2\right]dx - (2+z)xdz = 0;$$

finally,

$$zdx + (2+z)xdz = 0, \quad (1.3.31)$$

which is an **ODE with separable variables**.

Dividing by xz, we get

$$\frac{dx}{x} + \frac{2+z}{z}dz = 0, \quad (1.3.32)$$

which is an **ODE with separate variables**.

Its general solution is

$$\int \frac{dx}{x} + \int \left(1 + \frac{2}{z}\right)dz = C,$$

or

$$\ln|x| + z + 2\ln|z| = C.$$

Turning back to the initial variables, we get

$$\ln|x| + \frac{y}{x} + 2\ln\left|\frac{y}{x}\right| = C.$$

Passing to exponentials, we obtain the general solution of the homogeneous ODE (1.3.28)

$$x \cdot \frac{y^2}{x^2} \cdot e^{\frac{y}{x}} = C, \qquad (1.3.33)$$

or

$$\boxed{y^2 = Cx e^{-\frac{y}{x}}}. \qquad (1.3.34)$$

1.3.4. EXACT/TOTAL DIFFERENTIAL EQUATIONS

They are of the form

$$P(x,y)\,dx + Q(x,y)\,dy = 0. \qquad (1.3.35)$$

Definition 1.4. A first-order ODE is said to be an *exact* or *total differential equation* if there exists a differentiable function $F = F(x,y)$, such that $dF \equiv P(x,y)\,dx + Q(x,y)\,dy$.

From the study of differentials (see e.g. [2][3][5][10]) it is known that:

$$dF \equiv P(x,y)\,dx + Q(x,y)\,dy \text{ if and only if}$$

$$\frac{\partial P}{\partial y} = \frac{\partial Q}{\partial x}. \qquad (1.3.36)$$

COROLLARY:

The general solution of an exact/total differential equation is

$$F(x,y) = C, \qquad (1.3.37)$$

where C is an arbitrary constant.

Therefore the solving of an exact/total differential equation is reduced to finding a function of two variables, given its differential.

METHOD OF SOLVING

STEP 1. We compute the partial derivatives $\dfrac{\partial P}{\partial y}, \dfrac{\partial Q}{\partial x}$; if they coincide, then the ODE is an exact/total differential ODE, namely there exists F such that

$$dF \equiv P(x,y)dx + Q(x,y)dy.$$

STEP 2. Because the differential of a function is (see e.g. [2][3][5][10])

$$dF = \frac{\partial F}{\partial x}dx + \frac{\partial F}{\partial y}dy, \qquad (1.3.38)$$

hence

$$\begin{cases} \dfrac{\partial F}{\partial x} = P, \\ \dfrac{\partial F}{\partial y} = Q. \end{cases} \qquad (1.3.39)$$

By integrating the first relation with respect to x, we can shape F as:

$$F(x,y) = \int_{x_0}^{x} P(t,y)\,dt + \varphi(y), \qquad (1.3.40)$$

where φ is an arbitrary function, which depends only on y.

By differentiating both members of this relation with respect to y, we get

$$\frac{\partial F}{\partial y} = \int_{x_0}^{x} \frac{\partial P}{\partial y}(t, y)\, dt + \varphi'(y), \qquad (1.3.41)$$

where x_0 is fixed, but arbitrarily chosen, so that (x_0, y) would belong to the domain of definition of P and Q.

Taking into account the condition (1.3.36), we infer

$$\begin{aligned}\frac{\partial F}{\partial y} &= \int_{x_0}^{x} \frac{\partial Q}{\partial t}(t, y)\, dt + \varphi'(y) = \\ &= Q(x, y) - Q(x_0, y) + \varphi'(y).\end{aligned} \qquad (1.3.42)$$

By comparing this relation to the expression of $\dfrac{\partial F}{\partial y}$ from (1.3.39), it follows that

$$Q(x, y) - Q(x_0, y) + \varphi'(y) = Q(x, y), \qquad (1.3.43)$$

whence

$$\varphi'(y) = Q(x_0, y), \qquad (1.3.44)$$

thus the expression of φ is

$$\varphi(y) = \int_{y_0}^{y} Q(x_0, t)\, dt, \qquad (1.3.45)$$

where y_0 was chosen under the same assumptions as x_0.

Finally, we find for F

$$F(x,y) = \int_{x_0}^{x} P(t,y)\,dt + \int_{y_0}^{y} Q(x_0,t)\,dt, \qquad (1.3.46)$$

so that the **general solution of the exact/total differential ODE** is of the form

$$\boxed{\int_{x_0}^{x} P(t,y)\,dt + \int_{y_0}^{y} Q(x_0,t)\,dt = C}, \qquad (1.3.47)$$

whith C an arbitrary constant.

Note that if we choose to integrate in the first place the second relation (1.3.39) with respect to y, we obtain the general solution under an equivalent to (1.3.47) form

$$\boxed{\int_{x_0}^{x} P(t,y_0)\,dt + \int_{y_0}^{y} Q(x,t)\,dt = C}. \qquad (1.3.48)$$

Example. Find the general solution of the equation

$$y\mathrm{e}^x\,dx + (y + \mathrm{e}^x)\,dy = 0. \qquad (1.3.49)$$

Solution.

I. We first verify if the condition (1.3.36) is fulfilled. We have

$$\begin{cases} P(x,y) = y\mathrm{e}^x, \\ Q(x,y) = y + \mathrm{e}^x, \end{cases}$$

thus

$$\begin{cases} \dfrac{\partial P(x,y)}{\partial y} = \mathrm{e}^x, \\ \dfrac{\partial Q(x,y)}{\partial x} = \mathrm{e}^x, \end{cases}$$

and it follows that (1.3.49) is an *exact/total differential equation*.

II. This means that *there exists a function F of class* C^1 *such that*

$$\begin{cases} \dfrac{\partial F}{\partial x} = ye^x, \\ \dfrac{\partial F}{\partial y} = y + e^x. \end{cases} \quad (1.3.50)$$

From the first relation (1.3.50) we infer

$$F(x, y) = ye^x + \varphi(y), \quad (1.3.51)$$

whence

$$\dfrac{\partial F}{\partial y}(x, y) = e^x + \varphi'(y). \quad (1.3.52)$$

By equating this expression with the corresponding one from (1.3.50), we obtain

$$e^x + \varphi'(y) = e^x + y, \quad (1.3.53)$$

which yields

$$\varphi'(y) = y \;\rightarrow\; \varphi(y) = \dfrac{y^2}{2}. \quad (1.3.54)$$

By replacing this expression in (1.3.51), we obtain **the general solution of the ODE** (1.3.49):

$$\boxed{ye^x + \dfrac{y^2}{2} = C}. \quad (1.3.55)$$

Remarks. The direct application of the above-mentioned general formulas leads to the same result.

a) We apply the formula (1.3.47), and we use the conditions $x_0 = 0$, $y_0 = 0$.

We obtain

$$F(x,y) = \int_0^x P(t,y)\,dt + \int_0^y Q(x_0,t)\,dt = \int_0^x ye^t\,dt + \int_0^y (t + e^0)\,dt =$$

$$= ye^t\Big|_{t=0}^{t=x} + \left(\frac{t^2}{2} + t\right)\Big|_{t=0}^{t=y} = ye^x - y + \frac{y^2}{2} + y,$$

therefore the general solution of the ODE is again

$$\boxed{ye^x + \frac{y^2}{2} = C},$$

with the arbitrary constant C.

b) We now apply the formula (1.3.48); here, we can also use the same conditions $x_0 = 0$, $y_0 = 0$. We obtain

$$F(x,y) = \int_0^x P(t, y_0)\,dt + \int_0^y Q(x,t)\,dt = \int_0^x 0 \cdot e^t\,dt + \int_0^y (t + e^x)\,dt$$

$$= \left(te^x + \frac{t^2}{2}\right)\Big|_{t=0}^{t=y},$$

therefore the general solution of the ODE is the same as before

$$\boxed{ye^x + \frac{y^2}{2} = C},$$

with the arbitrary constant C.

1.3.5. INTEGRATING FACTOR

We saw that the method of solving an exact/total differential equation is extremely simple. This tempted the scientists to find ways in order to apply this attractive idea in other situations.

Consider the equation

$$P(x,y)dx + Q(x,y)dy = 0. \qquad (1.3.56)$$

Suppose that this is not an exact/total differential equation.

Let us put the following

QUESTION. *Is it possible to find a multiplicative factor* $\mu = \mu(x,y)$, *which transforms* (1.3.56) *into an exact/total differential equation?*

The function $\mu(x,y)$ is called an ***integrating/integrant factor.***

We can easily prove that:

There always exists an integrating factor.

A first-order ODE admits infinitely many integrating factors.

Any integrating factor of a first-order ODE is of the form $\varphi(U)\mu(x,y)$, where $U(x,y) = C$ is an integral (or a solution) of the equation, and $\mu(x,y)$ is an integrating factor.

Given two integrating factors of a first-order ODE, its solution can be expressed without quadratures.

HOW DO WE FIND THE INTEGRATING FACTOR?

Suppose that the problem is solved; therefore we already multiplied the equation (1.3.56) by a function $\mu = \mu(x,y)$ and we obtained

$$(\mu P)dx + (\mu Q)dy = 0, \qquad (1.3.57)$$

which is an exact/total differential ODE. According to the properties of a differential (see e.g. [2][3][5][10]), there exists a function $F = F(x, y)$, of class C^1 such that

$$dF(x,y) \equiv (\mu P)dx + (\mu Q)dy, \qquad (1.3.58)$$

which implies

$$\frac{\partial F}{\partial x} = \mu P, \quad \frac{\partial F}{\partial y} = \mu Q. \qquad (1.3.59)$$

If F is of class C^2, then, obviously,

$$\frac{\partial}{\partial y}(\mu P) = \frac{\partial}{\partial x}(\mu Q), \qquad (1.3.60)$$

because its mixed derivatives coincide, according to the Schwartz theorem (see e.g. [2][3][5][10]).

By differentiation, we obtain

$$\mu \frac{\partial P}{\partial y} + P \frac{\partial \mu}{\partial y} = \mu \frac{\partial Q}{\partial x} + Q \frac{\partial \mu}{\partial x}, \qquad (1.3.61)$$

which is, in fact, a partial differential equation satisfied by μ; this issue seems to be more complicated than the ODE (1.3.56) itself!

Actually, we don't need the general solution of the PDE (1.3.61). We can therefore suppose that $\mu = \mu(\omega)$, where $\omega = \omega(x, y)$ is a known function which depends on x and y. Since μ depends on x and y only through ω, we can apply the chain rule:

$$\frac{\partial \mu}{\partial x} = \frac{d\mu}{d\omega} \cdot \frac{\partial \omega}{\partial x}, \quad \frac{\partial \mu}{\partial y} = \frac{d\mu}{d\omega} \cdot \frac{\partial \omega}{\partial y}. \quad (1.3.62)$$

Introducing these expressions in (1.3.61), we obtain

$$\frac{d\mu}{d\omega}\left(P\frac{\partial \omega}{\partial y} - Q\frac{\partial \omega}{\partial x}\right) = \mu\left(\frac{\partial Q}{\partial x} - \frac{\partial P}{\partial y}\right), \quad (1.3.63)$$

or

$$\frac{d\mu}{d\omega} = \underbrace{\frac{\dfrac{\partial Q}{\partial x} - \dfrac{\partial P}{\partial y}}{P\dfrac{\partial \omega}{\partial y} - Q\dfrac{\partial \omega}{\partial x}}}_{\varphi(\omega)} \cdot \mu. \quad (1.3.64)$$

If the new expression, denoted by $\varphi(\omega)$, is a function which depends only on ω, then the equation (1.3.64) can be written as

$$d\mu - \varphi(\omega) \cdot \mu \, d\omega = 0. \quad (1.3.65)$$

This is **an *ODE with separable variables*,** in μ and ω.

Dividing by μ, we infer

$$\frac{d\mu}{\mu} = \varphi(\omega) d\omega, \quad (1.3.66)$$

with the general solution

$$\ln|\mu| = \int \varphi(\omega) d\omega + \ln C, \quad (1.3.67)$$

wherefrom

$$\mu = C \cdot e^{\int \varphi(\omega) d\omega}. \quad (1.3.68)$$

In fact, we are interested only in a particular solution of the ODE (1.3.66); for example, we can take $C = 1$.

After having found the integrating factor, we multiply the given equation by it, thus obtaining an exact/total differential equation, which we solve according to the model from the previous paragraph.

Remark. This method of solving depends on the choice of the function ω; this choice depends itself on user's ability. However, in many cases, ω is given or it is of a simple form.

Example. Solve the ODE

$$\underbrace{\left(-2x + \frac{1}{x+y}\right)}_{P} dx + \underbrace{\left(-x + y + \frac{1}{x+y}\right)}_{Q} dy = 0. \qquad (1.3.69)$$

knowing that it allows an integrating factor of the form $\mu = \mu(x+y)$.

Solution. We compute

$$\frac{\partial P}{\partial y} = -\frac{1}{(x+y)^2}, \quad \frac{\partial Q}{\partial x} = -1 - \frac{1}{(x+y)^2}. \qquad (1.3.70)$$

This means that

$$\frac{\partial P}{\partial y} \neq \frac{\partial Q}{\partial x}, \qquad (1.3.71)$$

therefore (1.3.69) is not an exact/total differential ODE.

So, let us try to find an integrating factor of the form $\mu = \mu(\omega)$, where $\omega = x + y$, according to the hint. One must have

$$\frac{\partial}{\partial y}(\mu P) = \frac{\partial}{\partial x}(\mu Q). \qquad (1.3.72)$$

In this case,

$$\frac{\partial \mu}{\partial y} = \frac{d\mu}{d\omega} \cdot 1, \quad \frac{\partial \mu}{\partial x} = \frac{d\mu}{d\omega} \cdot 1.$$

On the other hand, the above-mentioned calculation leads to

$$\frac{\partial Q}{\partial x} - \frac{\partial P}{\partial y} = -1.$$

From (1.3.72) we obtain

$$\frac{d\mu}{d\omega} \cdot P + \mu \frac{\partial P}{\partial y} = \frac{d\mu}{d\omega} \cdot Q + \mu \frac{\partial Q}{\partial x},$$

or

$$\frac{d\mu}{d\omega} \cdot \underbrace{(P - Q)}_{\underbrace{-(x+y)}_{\omega}} = \mu \underbrace{\left(\frac{\partial P}{\partial y} - \frac{\partial Q}{\partial x}\right)}_{-1}. \qquad (1.3.73)$$

This means that

$$-\frac{d\mu}{d\omega} \cdot \omega = -\mu,$$

which is an equation with separable variables. Dividing by $\omega\mu$, we obtain the equation with **separate variables**

$$\frac{d\mu}{\mu} = \frac{d\omega}{\omega},$$

for which we find the following particular solution

$$\ln|\mu| = \ln|\omega|.$$

Thus, the integrating factor is $\mu = \omega$, or else

$$\boxed{\mu = x + y}.$$

We multiply the EDO by $(x+y)$, thus getting

$$\underbrace{\left[-2x(x+y)+1\right]}_{p} dx + \underbrace{\left(y^2 - x^2 + 1\right)}_{q} dy = 0. \qquad (1.3.74)$$

This is an exact/total differential equation, because

$$\frac{\partial p}{\partial y} = -2x = \frac{\partial q}{\partial x}.$$

Therefore, we must look for a function F such that

$$\begin{cases} \dfrac{\partial F}{\partial x} = p = -2x^2 - 2xy + 1 \\ \dfrac{\partial F}{\partial y} = q = y^2 - x^2 + 1 \end{cases} \Rightarrow F = \frac{y^3}{3} - x^2 y + y + \varphi(x).$$

Differentiating F with respect to x, we deduce:

$$\frac{\partial F}{\partial x} = -2xy + \varphi'(x).$$

One must have

$$-2xy + \varphi'(x) = -2x^2 - 2xy + 1,$$

wherefrom

$$\varphi(x) = -\frac{2}{3}x^3 + x.$$

Finally, the function F is given by

$$F(x,y) = \frac{y^3}{3} - x^2 y + y - \frac{2}{3}x^3 + x;$$

hence the general solution of the equation (1.3.69) is

$$\boxed{y^3 - 3x^2 y + 3y - 2x^3 + 3x = C}. \qquad (1.3.75)$$

1.3.6. FIRST ORDER LINEAR ODEs

The equation

$$y' + p(x)y = q(x), \quad p, q \in C^1(I), I \subseteq \Re, \qquad (1.3.76)$$

where $y' = \dfrac{dy}{dx}$, defines a ***first-order non-homogeneous linear ODE***.

The ***associated homogeneous equation*** is

$$y' + p(x)y = 0. \qquad (1.3.77)$$

A. The left-hand side of the equation (1.3.76) defines ***the operator*** L, which associates the function $y' + p(x)y$ to each one of the functions y, namely

$$Ly \equiv y' + p(x)y. \qquad (1.3.78)$$

For example, if L is defined as

$$Ly \equiv y' + 2y, \qquad (1.3.79)$$

then it realises the following function-to-function correspondence:

$$y_1 = x \xrightarrow{L} Ly_1 = 1 + 2x = 3x,$$
$$y_2 = \cos x \xrightarrow{L} Ly_2 = -\sin x + 2\cos x, \quad (1.3.80)$$
$$y_3 = e^{-2x} \xrightarrow{L} Ly_3 = -2e^{-2x} + 2e^{-2x} = 0.$$

One can affirm that the operator L expressed by (1.3.78) is defined as follows:

$$L : C^1(I) \to C^0(I). \quad (1.3.81)$$

B. *The operator L expressed by* (1.3.78) *is linear*.

Definition 1.5. An operator $L : X \to Y$, where X, Y are real/complex vector spaces, is called *linear* if

$$L(\alpha x_1 + \beta x_2) = \alpha L(x_1) + \beta L(x_2), \quad (1.3.82)$$

for any $x_1, x_2 \in X$ and any real/complex α, β.

If $y_1, y_2 \in C^0(I)$ and α, β are real/complex constants, then

$$\begin{aligned}
L(\alpha y_1 + \beta y_2) &\underset{\text{def}}{=} (\alpha y_1 + \beta y_2)' + p(x)(\alpha y_1 + \beta y_2) \\
&= \alpha y_1' + \beta y_2' + \alpha p(x) y_1 + \beta p(x) y_2 \\
&= \alpha \underbrace{\left[y_1' + p(x) y_1 \right]}_{Ly_1} + \beta \underbrace{\left[y_2' + p(x) y_2 \right]}_{Ly_2} = \\
&= \alpha L(y_1) + \beta L(y_2),
\end{aligned} \quad (1.3.83)$$

therefore L is *linear*, according to the above-mentioned definition.

Remark. **We identify a linear differential operator by that that always in its structure, the unknown function, as well as its derivative(s) of any order, are of first degree.**

Definition 1.6. The *kernel of an operator* $L : X \to Y$, denoted by "ker", is the subset of X whose elements are mapped by L into the null element of Y, i.e.

$$\ker L \equiv \{\, x \in X \mid L(x) = 0_Y \,\}. \qquad (1.3.84)$$

It is well-known (see e.g. [2][5]) that ker L is a vector subspace of X.

For the linear differential operator (1.3.78), the kernel is

$$\ker L \equiv \{\, y \in C^1(I) \mid Ly = 0 \,\}, \qquad (1.3.85)$$

thus ker L coincides with the set of the solutions of the homogeneous linear ODE (1.3.77).

The linear and homogeneous equation (1.3.78) can be written as an equation with separable variables:

$$\frac{dy}{dx} + p(x) y = 0 \quad \Rightarrow \quad dy + p(x) y \, dx = 0, \qquad (1.3.86)$$

whence, dividing by y, we get step by step

$$\begin{aligned}
\frac{dy}{y} &= -p(x) \, dx, \\
d(\ln y) &= -p(x) \, dx, \\
\ln y &= -\int p(x) \, dx + c.
\end{aligned} \qquad (1.3.87)$$

In the last expression, c is an arbitrary constant and it can be taken of the form $\ln C$. Passing to exponentials in the last equality, we get

$$y = C e^{-\int p(x) \, dx}, \qquad (1.3.88)$$

which is **the general solution of the homogeneous ODE** (1.3.77).

Remark. Formula (1.3.88) shows the vector subspace $\ker L$ is of dimension 1.

From now on, we shall write the ODE (1.3.76) in the form

$$Ly \equiv y' + p(x)y = q(x). \qquad (1.3.89)$$

We can easily prove that

Theorem 1.1. *The general solution of the non-homogeneous ODE (1.3.89) is the sum between a particular solution of the non-homogeneous ODE and the general solution of the associated homogeneous ODE.*

Proof. Let Y be a particular solution of the non-homogeneous ODE (1.3.89). This means that

$$LY \equiv Y' + p(x)Y = q(x). \qquad (1.3.90)$$

Let us consider the change of function

$$y = Y + z. \qquad (1.3.91)$$

Introducing it in (1.3.89), we get

$$Ly = L(Y+z) \underset{L\,\text{linear}}{=} LY + Lz = q(x) + Lz. \qquad (1.3.92)$$

But $Ly = q(x)$. From (1.3.92) it follows that $Lz = 0$, which means that $z \in \ker L$. ∎

HOW DO WE FIND Y?

The answer to this question is given by

The method of variation of arbitrary constants (also called **Lagrange's method**, or **variation of parameters**).

The idea of the method is to look for Y in the form (1.3.88), but instead of considering C as a constant, we take it as a function of x:

$$Y = C(x)e^{-\int p(x)dx}. \qquad (1.3.93)$$

Then

$$Y' = C'(x)e^{-\int p(x)dx} - p(x)C(x)e^{-\int p(x)dx}; \qquad (1.3.94)$$

replacing this into the non-homogeneous equation (1.3.89), we get

$$\begin{aligned} LY \equiv Y' + p(x)Y &= C'(x)e^{-\int p(x)dx} - p(x)C(x)e^{-\int p(x)dx} \\ &\quad + p(x)C(x)e^{-\int p(x)dx} = C'(x)e^{-\int p(x)dx}. \end{aligned} \qquad (1.3.95)$$

But $LY = q(x)$, hence

$$C'(x)e^{-\int p(x)dx} = q(x), \qquad (1.3.96)$$

which leads to

$$C'(x) = q(x)e^{\int p(x)dx}, \qquad (1.3.97)$$

Hence C is obtained by integration:

$$C(x) = \int q(x)e^{\int p(x)dx}\,dx. \qquad (1.3.98)$$

Finally, the particular solution Y is straightforwardly obtained by quadratures

$$Y(x) = e^{-\int p(x)dx}\int q(x)e^{\int p(x)dx}\,dx. \qquad (1.3.99)$$

Taking into account the theorem 1.1, it follows that

The general solution of the non-homogeneous linear ODE is obtained by quadratures and it is given by

$$y(x) = Ce^{-\int p(x)dx} + e^{-\int p(x)dx}\int q(x)e^{\int p(x)dx}\,dx, \qquad (1.3.100)$$

or, equally, by

$$y(x) = e^{-\int p(x)dx}\left(C + \int q(x)e^{\int p(x)dx}\,dx\right), \qquad (1.3.101)$$

where C is an arbitrary constant.

In order to solve a first-order linear ODE, we can use one of the two last formulas, but, in practice, it is much easier to proceed directly. The above computation emphasizes the following

METHOD OF SOLVING

STEP 1.

We consider the associated to (1.3.78) homogeneous equation:

$$Lz \equiv z' + p(x)z = 0. \qquad (1.3.102)$$

We have already proved that the general solution of this equation is given by formula (1.3.88), therefore

$$z = Ce^{-\int p(x)dx}. \qquad (1.3.103)$$

STEP 2.

According to the theorem 1.1, we must find Y – a particular solution of the ODE (1.3.78).

This solution is obtained by Lagrange's method (variation of arbitrary constants), as previously shown.

Example. Find the general solution of the following ODE:

a) $y' + y = 0$.

Solution. It is a first-order ODE. The unknown function y and its derivative are of first degree, therefore the ODE is linear. As the free term is missing, the ODE is homogeneous. We have, step by step

$$\frac{dy}{dx} + y = 0,$$
$$dy + y\,dx = 0,$$
$$\frac{dy}{y} + dx = 0;$$

the last one is an equation with separate variables. According to formula (1.3.4), its general solution is

$$\int \frac{dy}{y} = -\int dx + \ln C,$$
$$\ln|y| = -x + \ln C,$$

where C is an arbitrary constant. Passing to exponentials, we get

$$\boxed{y = Ce^{-x}.}$$

b) $y' + xy = 0$.

Solution. It is a first-order homogeneous linear ODE. Step by step, we get

$$\frac{dy}{dx} + xy = 0,$$
$$dy + xy\,dx = 0,$$
$$\frac{dy}{y} + x\,dx = 0;$$

the last one is an equation with separate variables, whose general solution is

$$\int \frac{dy}{y} = -\int x\,dx + \ln C,$$

or

$$\ln|y| = -\frac{x^2}{2} + \ln C,$$

with C an arbitrary constant. Passing to exponentials, it results

$$\boxed{y = Ce^{-\frac{x}{2}}}.$$

c) $y' + y = e^x$.

Solution. It is a first-order non-homogeneous linear ODE.

STEP 1. The associated homogeneous equation is $z' + z = 0$. Its general solution has already been found at the example a). It is

$$z = Ce^{-x}.$$

STEP 2. In order to find a particular solution Y for the non-homogeneous equation, we use *the method of variation of parameters,* searching for Y of the form

$$Y = C(x)e^{-x}.$$

Introducing it in the non-homogeneous equation, we have:

$$\begin{array}{l} 1 | Y = C(x)e^{-x} \\ 1 | Y' = C'(x)e^{-x} - C(x)e^{-x} \end{array} \Bigg\| +$$

$$Y' + Y = C'(x)e^{-x} - C(x)e^{-x} + C(x)e^{-x} = C'(x)e^{-x},$$

and, as

$$Y' + Y = e^x,$$

we deduce

$$C'(x)e^{-x} = e^x,$$

i.e.,

$$C'(x) = e^{2x},$$

whence

$$C(x) = \frac{1}{2}e^{2x}.$$

Then

$$Y = \frac{1}{2}e^{2x}e^{-x} = \frac{1}{2}e^x.$$

The general solution of the non-homogeneous ODE is $y = z + Y$, therefore

$$\boxed{y = Ce^{-x} + \frac{1}{2}e^x}.$$

d) $y' + xy = xe^{-\frac{x^2}{2}}$.

Solution. It is a first-order non-homogeneous linear ODE.

STEP 1. The associated homogeneous equation is

$$z' + xz = 0.$$

Its general solution has already been found at example **b)**. It is $z = Ce^{-\frac{x^2}{2}}$.

STEP 2. In order to find a particular solution Y of the non-homogeneous equation, we use Lagrange's method.

We search for Y of the form $Y = C(x)e^{-\frac{x^2}{2}}$; introducing it in the non-homogeneous equation, it follows

$$x \left|\begin{array}{l} Y = C(x)e^{-\frac{x^2}{2}} \\ Y' = C'(x)e^{-\frac{x^2}{2}} - xC(x)e^{-\frac{x^2}{2}} \end{array}\right| +$$

$$Y' + xY = C'(x)e^{-\frac{x^2}{2}} - xC(x)e^{-\frac{x^2}{2}} + xC(x)e^{-\frac{x^2}{2}} = C'(x)e^{-\frac{x^2}{2}},$$

and since

$$Y' + xY = xe^{-\frac{x^2}{2}},$$

we get

$$C'(x)e^{-\frac{x^2}{2}} = xe^{-\frac{x^2}{2}},$$

i.e., $C'(x) = x$, hence

$$C(x) = \frac{x^2}{2}.$$

We obtain $Y = \frac{x^2}{2} e^{-\frac{x^2}{2}}$.

The general solution of the non-homogeneous ODE is $y = z + Y$, therefore

$$y = Ce^{-\frac{x^2}{2}} + \frac{x^2}{2} e^{-\frac{x^2}{2}},$$

or

$$\boxed{y = \left(C + \frac{x^2}{2}\right) e^{-\frac{x^2}{2}},}$$

where C is an arbitrary constant.

1.3.7. BERNOULLI'S EQUATION

It is of the form

$$y' + p(x)y = q(x)y^{\alpha},$$
$$\alpha \notin \{0,1\}, \; p,q \in C^{0}(I), \; I \subseteq \mathfrak{R}. \tag{1.3.104}$$

➢ If $\alpha = 0$, we obtain the first-order non-homogeneous linear ODE $y' + py = q$.

➢ If $\alpha = 1$, we obtain the first-order homogeneous linear ODE $y' + (p-q)y = 0$.

METHOD OF SOLVING

We ***divide*** (1.3.104) by y^{α}:

$$\frac{y'}{y^{\alpha}} + p(x) \cdot \frac{1}{y^{\alpha-1}} = q(x). \tag{1.3.105}$$

We differentiate $\dfrac{1}{y^{\alpha-1}} = y^{1-\alpha}$:

$$\frac{d}{dx}(y^{1-\alpha}) = (1-\alpha)y^{-\alpha}y' = (1-\alpha)\frac{y'}{y^{\alpha}}. \tag{1.3.106}$$

Therefore (1.3.104) turns into

$$\frac{1}{1-\alpha}\frac{d}{dx}\left(\frac{1}{y^{\alpha-1}}\right) + p(x)\frac{1}{y^{\alpha-1}} = q(x). \tag{1.3.107}$$

We now denote by

$$u = \frac{1}{y^{\alpha-1}}, \qquad (1.3.108)$$

and we obtain

$$\frac{1}{1-\alpha}u' + p(x)u = q(x), \qquad (1.3.109)$$

which is a **non-homogeneous linear ODE**, with u as unknown function. After finding the solution of this linear equation, we get back to y by (1.3.108).

Example. Solve the equation

$$y' + \frac{2}{3x}y = \frac{1}{3}y^2. \qquad (1.3.110)$$

We identify it as a Bernoulli equation, with $\alpha = 2$.

Solution.

♣ *We divide the equation by* y^2:

$$\frac{y'}{y^2} + \frac{2}{3x} \cdot \frac{1}{y} = \frac{1}{3}.$$

We use the notation $u = \dfrac{1}{y} \Rightarrow u' = -\dfrac{y'}{y^2}$, thus getting

$$-u' + \frac{2}{3x}u = \frac{1}{3}, \qquad (1.3.111)$$

which is a **non-homogeneous linear ODE**.

♣ *We solve the linear equation* (1.3.111).

• The associated homogeneous ODE is

$$-u' + \frac{2}{3x}u = 0. \qquad (1.3.112)$$

We obtain $\dfrac{u'}{u} = \dfrac{2}{3x}$, hence $\ln|u| = \dfrac{2}{3}\ln|x| + \ln|C|$.

The general solution of the homogeneous ODE (1.3.112) is thus

$$u = C \cdot x^{\frac{2}{3}}. \qquad (1.3.113)$$

- Therefore the general solution of the non-homogeneous equation (1.3.111) is

$$u = C \cdot x^{\frac{2}{3}} + U, \qquad (1.3.114)$$

where U is a particular solution of (1.3.111), which can be obtained by *variation of parameters*. We successively deduce

$$\left. \begin{array}{l} U(x) = C(x) \cdot x^{\frac{2}{3}} \\ U'(x) = C'(x) \cdot x^{\frac{2}{3}} + \dfrac{2}{3} \cdot x^{-\frac{1}{3}} \cdot C(x) \end{array} \right| \begin{array}{c} \frac{2}{3x} \\ -1 \end{array}$$

$$-U' + \dfrac{2}{3x} U = -C'(x) \cdot x^{\frac{2}{3}} = \dfrac{1}{3},$$

hence

$$C'(x) = -\dfrac{1}{3} \cdot x^{-\frac{2}{3}},$$

i.e.,

$$C(x) = -\dfrac{1}{3}\left(1 - \dfrac{2}{3}\right)^{-1} \cdot x^{1-\frac{2}{3}} = -x^{\frac{1}{3}};$$

consequently, the particular solution U is

$$U = -x.$$

The general solution of the ODE (1.3.111) is then

$$u = -x + C \cdot x^{\frac{2}{3}}.$$

♣ **The general solution of the Bernoulli equation (1.3.110) is given by**

$$y = \left(-x + C \cdot x^{\frac{2}{3}}\right)^{-1}.$$

1.3.8. RICCATI'S EQUATION

It is of the form

$$y' + p(x)y + q(x)y^2 = r(x), \ p,q,r \in C^0(I), \ I \subseteq \Re. \quad (1.3.115)$$

➤ For $q = 0$, we get the non-homogeneous linear ODE $y' + p(x)y = r(x)$.

➤ For $r = 0$, we get the Bernoulli equation $y' + p(x)y = -q(x)y^2$.

Given a particular solution $Y(x)$, the Riccati equation is solved by quadratures.

METHOD OF SOLVING

Indeed, by putting

$$y = Y + z, \quad (1.3.116)$$

we have $y' = Y' + z'$; replacing it in (1.3.115), we deduce

$$Y' + z' + p(x)(Y + z) + q(x)(Y^2 + 2Yz + z^2) = r(x). \quad (1.3.117)$$

But $Y' + pY + qY^2 = r(x)$, therefore Z satisfies

$$z' + \left[p(x) + 2q(x)Y \right] z + q(x) z^2 = 0, \qquad (1.3.118)$$

which is a Bernoulli equation, with $\alpha = 2$. After solving it, we get back to y, with the change of function (1.3.116).

Example. Solve the equation

$$y' = \frac{1}{3} y^2 + \frac{2}{3x^2}, \qquad (1.3.119)$$

knowing that it allows the particular solution $Y(x) = -\frac{1}{x}$.

Solution.

By using the change of function

$$y = -\frac{1}{x} + z, \qquad (1.3.120)$$

we obtain

$$\frac{1}{x^2} + z' = \frac{1}{3} \left(\frac{1}{x^2} - 2\frac{z}{x} + z^2 \right) + \frac{2}{3x^2}.$$

This leads to the Bernoulli equation

$$z' + \frac{2}{3x} z = \frac{1}{3} z^2, \qquad (1.3.121)$$

which is precisely the above example given at the previous section 1.3.7. For this equation, we found the general solution

$$z = \left(-x + C \cdot x^{\frac{2}{3}} \right)^{-1}.$$

We now get back to y, with the change of function (1.3.120).

It follows that the general solution of Riccati's equation (1.3.119) is

$$\boxed{y = -\frac{1}{x} + \left(-x + C \cdot x^{\frac{2}{3}}\right)^{-1}},$$

with C an arbitrary constant.

Comment. The Ricatti equation is very important in applications in mechanics, engineering, physics, chemistry, etc.; this is why it has been thoroughly analysed. It has a series of special properties (for example, any four distinct solutions of a given Riccati equation are always in a cross-ratio). The Riccati systems of equations are amongst the most used ones in modern researches of natural science.

1.3.9. CLAIRAUT'S EQUATION

It is of the form

$$y = xy' + \varphi(y'). \tag{1.3.122}$$

METHOD OF SOLVING

We use the notation

$$y' = \frac{dy}{dx} = p, \tag{1.3.123}$$

whence we obtain

$$dy = p\,dx. \tag{1.3.124}$$

On the other hand, from (1.3.122) we get

$$y = xp + \varphi(p); \tag{1.3.125}$$

by differentiation, it follows

$$\underbrace{dy}_{p\,dx} = p\,dx + x\,dp + \varphi'(p)dp. \tag{1.3.126}$$

Equating the two expression of dy, we get

$$p\,dx = p\,dx + x\,dp + \varphi'(p)dp, \tag{1.3.127}$$

therefore

$$(x + \varphi'(p))dp = 0. \tag{1.3.128}$$

It follows that at least one of the following equalities is true

$$\begin{cases} dp = 0, \\ x + \varphi'(p) = 0. \end{cases} \tag{1.3.129}$$

- **Case a).** If $dp = 0$, then $p = C$, hence

$$\boxed{y = xC + \varphi(C)}, \tag{1.3.130}$$

where C is an arbitrary constant. The relation (1.3.130) is **the general solution of Clairaut's equation.** From the geometric point of view, the solution of the Clairaut equation is a **pencil of straight lines.**

- **Case b).** If $x + \varphi'(p) = 0$, then $x = -\varphi'(p)$. Thus

$$y = -\varphi'(p) \cdot p + \varphi(p).$$

It follows that

$$\begin{cases} x = -\varphi'(p), \\ y = -\varphi'(p) \cdot p + \varphi(p), \end{cases} \quad (1.3.131)$$

which represent the parametric equations of an integral curve for the Clairaut equation. This is **not** obtained from the general solution, by giving particular values to C. Consequently, this solution is a **singular solution**. From the geometrical point of view, it is **the envelope of the pencil of straight lines** represented by the general solution.

Indeed, if $F(x,y,C) = 0$ is a sheaf of curves, then, by eliminating C between the relations

$$\begin{cases} F(x,y,C) = 0, \\ \dfrac{\partial F}{\partial C}(x,y,C) = 0, \end{cases} \quad (1.3.132)$$

we obtain **the sheaf envelope**.

In the case of Clairaut's equation, F and $\dfrac{\partial F}{\partial C}$ are of the form

$$\begin{cases} F(x,y,C) \equiv xC + \varphi(C) - y = 0, \\ \dfrac{\partial F}{\partial C}(x,y,C) \equiv x + \varphi'(C) = 0. \end{cases} \quad (1.3.133)$$

By eliminating C between these two equations, we get

$$\begin{cases} x = -\varphi'(C), \\ y = -\varphi'(C) \cdot C + \varphi(C), \end{cases} \quad (1.3.134)$$

which are, actually, the parametric equations of the singular solution.

Therefore, the singular solution of the Clairaut equation is the envelope of the pencil of straight lines that represents geometrically its general solution.

Example. Solve the equation

$$y = xy' - y'^2. \qquad (1.3.135)$$

Solution.

$$\left. \begin{array}{l} y' = p \\ y = x \cdot p - p^2 \end{array} \right| \Rightarrow \begin{cases} dy = p\,dx, \\ dy = p\,dx + x\,dp - 2p\,dp. \end{cases}$$

Equating the expressions of dy, we infer

$$p\,dx + x\,dp - 2p\,dp = p\,dx \quad \Rightarrow \quad (x - 2p)\,dp = 0,$$

i.e.,

$$\begin{cases} dp = 0, \\ x = 2p. \end{cases} \qquad (1.3.136)$$

- **Case a).** $dp = 0 \Rightarrow p = C$, therefore

$$\boxed{y = x \cdot C - C^2}, \qquad (1.3.137)$$

which represents the **general solution of Clairaut's equation.** We see that, geometrically, this is a pencil of straight lines.

- **Case b).** We have

$$\begin{cases} x = 2p, \\ y = 2p \cdot p - p^2 = p^2, \end{cases}$$

whence we get

$$\boxed{y = \frac{x^2}{4}}; \qquad (1.3.138)$$

this is *the singular solution of Clairaut's equation*, a parabola.

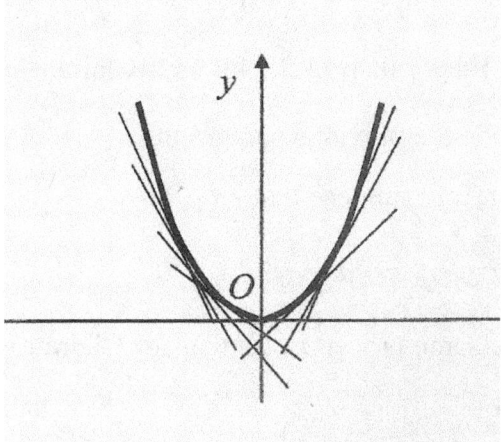

In the above figure, we see that the singular solution (the parabola) is tangent at each point to one of straight lines of the pencil (1.3.137), pencil which gives the general solution of Clairaut's equation (1.3.135).

1.3.10. LAGRANGE'S EQUATION

It is of the form

$$A(y')y + B(y')x + C(y') = 0, \qquad (1.3.139)$$

it is seen that it depends linearly on x and y. If $A(y') \neq 0$, dividing by it, we obtain

$$y = f(y')x + g(y'), \quad f(y') = -\frac{B(y')}{A(y')},$$
$$g(y') = -\frac{C(y')}{A(y')}. \qquad (1.3.140)$$

If $f(y') \equiv y'$, then (1.3.140) is a Clairaut equation; it has been presented in the previous paragraph.

We therefore suppose that $f(y') \neq y'$.

METHOD OF SOLVING

It is the same procedure as that for Clairaut's equation. We use the change

$$y' = \frac{dy}{dx} = p, \qquad (1.3.141)$$

whence we obtain that

$$dy = p\,dx. \qquad (1.3.142)$$

On the other hand, from (1.3.140) we get

$$y = xf(p) + g(p), \qquad (1.3.143)$$

which becomes after differentiation

$$\underset{pdx}{dy} = f(p)\,dx + xf'(p)\,dp + g'(p)\,dp. \qquad (1.3.144)$$

Equating the two expressions of dy, we get

$$p\,dx = f(p)\,dx + xf'(p)\,dp + g'(p)\,dp, \qquad (1.3.145)$$

therefore

$$[f(p)-p]\,\mathrm{d}x+[xf'(p)+g'(p)]\,\mathrm{d}p=0. \qquad (1.3.146)$$

If $f(p)=\mathrm{const}$, then (1.3.146) is an equation with separable variables and it can be solved as shown in section 1.3.2.

If $f(p)$ is not constant, then there are two possibilities:

a) $f(p)\neq p$. Then we divide (1.3.146) by $f(p)-p$, thus obtaining

$$\frac{\mathrm{d}x}{\mathrm{d}p}+\frac{f'(p)}{f(p)-p}x+\frac{g'(p)}{f(p)-p}=0. \qquad (1.3.147)$$

This is a non-homogeneous linear ODE, having x as unknown function and p as independent variable. After having solved it following the method described in paragraph 1.3.6, we obtain the solution in the form $x(p)=a_1(p)C+b_1(p)$, where C is an arbitrary constant.

From (1.3.143) it follows that

$$y(p)=[a_1(p)C+b_1(p)]f(p)+g(p), \qquad (1.3.148)$$

or

$$y(p)=a_2(p)C+b_2(p), \qquad (1.3.149)$$

where there were used the notations

$$a_2(p)=a_1(p)f(p),\quad b_2(p)=b_1(p)f(p)+g(p). \qquad (1.3.150)$$

Finally, we obtain **the general solution of the Lagrange equation** in parametric form

$$\begin{cases} x(p) = a_1(p)C + b_1(p), \\ y(p) = a_2(p)C + b_2(p). \end{cases} \quad (1.3.151)$$

b) If $f(p) - p = 0$ allows the real solutions p_i, by replacing them in the equation (1.3.140) and taking into account that $f(p_i) = p_i$, we get the solutions

$$y = x\, p_i + g(p_i), \quad (1.3.152)$$

which are equations of straight lines, for each p_i.

These solutions can be singular.

Example. Find the general solution of the equation

$$y = x - \frac{4}{9} y'^2 + \frac{8}{27} y'^3. \quad (1.3.153)$$

Solution. It is a Lagrange equation. Therefore we use the change

$$y' = p \;\rightarrow\; dy = p\, dx, \quad (1.3.154)$$

whence

$$y = x - \frac{4}{9} p^2 + \frac{8}{27} p^3, \quad (1.3.155)$$

which, by differentiation, leads to

$$dy = dx - \frac{8}{9} p\, dp + \frac{8}{9} p^2 dp. \quad (1.3.156)$$

Equating the two expressions of dy, we get

$$p\, dx = dx - \frac{8}{9} p\, dp + \frac{8}{9} p^2 dp, \quad (1.3.157)$$

or, after some elementary computation,

$$(p-1)\left[\,\mathrm{d}x - \frac{8}{9}p\,\mathrm{d}p\,\right] = 0. \qquad (1.3.158)$$

Therefore, at least one of the following equalities holds true:

$$\begin{cases} \mathrm{d}x - \dfrac{8}{9}p\,\mathrm{d}p = 0, \\ p = 1. \end{cases} \qquad (1.3.159)$$

a) The first equality is, in fact, the equation with separate variables

$$\mathrm{d}x - \frac{8}{9}p\,\mathrm{d}p = 0,$$

whose general solution is

$$x = \frac{4}{9}p^2 + C. \qquad (1.3.160)$$

From (1.3.155) it follows that

$$y = \frac{8}{27}p^3 + C. \qquad (1.3.161)$$

Therefore, *the general solution of* Lagrange's equations is obtained in *parametric form*

$$\begin{cases} x = \dfrac{4}{9}p^2 + C, \\ y = \dfrac{8}{27}p^3 + C. \end{cases} \qquad (1.3.162)$$

By eliminating p between the two expressions from (1.3.162), we find *the general solution* in *implicit form*

$$\boxed{(x-C)^3 = (y-C)^2}. \qquad (1.3.163)$$

b) The second equality (1.3.159) implies that $p=1$, which, replaced in (1.3.155), leads to the ***singular solution*** of the Lagrange equation:

$$\boxed{y = x - \frac{4}{27}}. \qquad (1.3.164)$$

1.4. THE METHOD OF SUCCESSIVE APPROXIMATION

In the previous paragraph, we emphasized some types of first-order ODE which can be solved by quadratures. This leads to practical analytic formulas. However, these types of equations are not as frequently met in practice as it should be expected. This shortcoming could be balanced by some methods of approximating the solutions.

One of the commonly used methods is ***the method of successive approximations***, or ***Picard's method***. As this method is constructive, we shall describe it in the general frame of the existence and uniqueness of the solution of the Cauchy problem.

1.4.1. CAUCHY-PICARD'S THEOREM OF EXISTENCE AND UNIQUENESS

Theorem 1.2. *Consider the Cauchy problem*

$$\begin{cases} \dfrac{dy}{dx} = f(x,y), \\ y(x_0) = y_0. \end{cases} \qquad (1.4.1)$$

Suppose that f satisfies the following conditions:

i) $f \in C^0(\Omega)$,
$\Omega = \{(x,y) \in \Re^2, |x - x_0| \le a, |y - y_0| \le b\}$;

ii) f is Lipschitz with respect to y, which means that there exists a positive constant K such that

$$|f(x,Y) - f(x,Z)| \le K|Y - Z|, \forall (x,Y), (x,Z) \in \Omega. \qquad (1.4.2)$$

Then the Cauchy problem (1.4.1) allows a unique solution $y \in C^1(I)$, *where I is the interval* $I = (x_0 - h, x_0 + h)$, *its length 2h being computed as follows:*

$$h = \min\left\{a, \dfrac{b}{M}\right\}, \quad M = \sup_{(x,y) \in \Omega} |f(x,y)|. \qquad (1.4.3)$$

*Proof. *M* exists and it is finite, due to the continuity of *f* on Ω, which is bounded and closed. We firstly prove

THE EXISTENCE OF THE SOLUTION

By integrating the equation from (1.4.1) and taking into account the Cauchy condition, we notice that the problem (1.4.1) is equivalent to the integral equation

$$y(x) = y_0 + \int_{x_0}^{x} f(t, y(t)) dt. \qquad (1.4.4)$$

Its existence is constructive, by

THE METHOD OF SUCCESSIVE APPROXIMATIONS,

also known as **Picard's method**, who introduced it.

The method is constructive and has a high degree of applicability. It can be also used to other problems.

In this case, we consider the following sequence approximating the solution of problem (1.4.4):

$$y_1(x) = y_0 + \int_{x_0}^{x} f(t, y_0) dt,$$

$$y_2(x) = y_0 + \int_{x_0}^{x} f(t, y_1(t)) dt, \qquad (1.4.5)$$

$$\cdots\cdots\cdots\cdots\cdots\cdots\cdots\cdots\cdots\cdots\cdots\cdots\cdots$$

$$y_n(x) = y_0 + \int_{x_0}^{x} f(t, y_{n-1}(t)) dt, \quad n \in \mathfrak{N}.$$

One should follow several steps:

I. Prove that the sequence $\{y_n\}_{n \in \mathfrak{N}}$ **is well defined** and that the range of any of the functions y_n, $n \in \mathfrak{N}$ lays in the interval $[y_0 - b, y_0 + b]$, for any $x \in I$.

* Indeed

$$|y_1(x) - y_0| \le \left| \int_{x_0}^{x} f(t, y_0) dt \right| \le M |x - x_0| \le$$

$$\le Mh \frac{b}{M} = b, \quad x \in I. \qquad (1.4.6)$$

Using the complete induction, we assume that

$$|y_j(x) - y_0| \le b, \quad x \in I, \, j = \overline{1, n-1}, \qquad (1.4.7)$$

and, as previously, we get

$$\left|y_n(x) - y_0\right| \leq \left|\int_{x_0}^{x} f(t, y_{n-1}(t))\,dt\right| \leq M\left|x - x_0\right| \leq \qquad (1.4.8)$$

$$\leq Mh\frac{b}{M} = b, \quad x \in I.$$

II. We show that the sequence $\{y_n\}_{n\in\mathfrak{N}}$ is uniformly and absolutely convergent on I.

* To this end, we consider the series

$$S \equiv y_0 + (y_1 - y_0) + \ldots + (y_n - y_{n-1}) + \ldots, \qquad (1.4.9)$$

with the partial sums $y_0 + (y_1 - y_0) + \ldots + (y_n - y_{n-1}) \equiv y_n$. We get, step by step,

$$\left|y_1(x) - y_0\right| \leq \left|\int_{x_0}^{x} f(t, y_0)\,dt\right| \leq M\left|x - x_0\right|,$$

$$\left|y_2(x) - y_1(x)\right| \leq \left|\int_{x_0}^{x} |f(t, y_1(t)) - f(t, y_0)|\,dt\right| \qquad (1.4.10)$$

$$\leq MK\left|\int_{x_0}^{x} |y_1(t) - y_0|\,dt\right| \leq MK\frac{(x - x_0)^2}{2!}.$$

We use again the complete induction. If the inequality

$$\left|y_k(x) - y_{k-1}(x)\right| \leq MK^{k-1}\frac{(x - x_0)^k}{k!}$$

is valid for $k = \overline{1, n-1}$, then

$$\left|y_n(x)-y_{n-1}(x)\right| \leq \left|\int_{x_0}^{x}\left|f\left(t, y_{n-1}(t)\right)-f\left(t, y_{n-2}(t)\right)\right| dt\right| \qquad (1.4.11)$$

$$\leq K\left|\int_{x_0}^{x}\left|y_{n-1}(t)-y_{n-2}(t)\right| dt\right|,$$

and since

$$K\left|\int_{x_0}^{x}\left|y_{n-1}(t)-y_{n-2}(t)\right| dt\right| \qquad (1.4.12)$$

$$\leq K \cdot \frac{MK^{n-2}}{(n-1)!}\left|\int_{x_0}^{x}\left|t-x_0\right|^{n-1} dt\right| = MK^{n-1}\frac{\left|x-x_0\right|^{n}}{n!},$$

we conclude that each term of the series (1.4.9) can be majorized by

$$\left|y_n(x)-y_{n-1}(x)\right| \leq MK^{n-1}\frac{\left|x-x_0\right|^{n}}{n!} \leq MK^{n-1}\frac{h^n}{n!}, \quad x \in I. \qquad (1.4.13)$$

The numeric series with positive numbers

$$Mh + MK\frac{h^2}{2!} + \ldots + MK^{n-1}\frac{h^n}{n!} + \ldots \qquad (1.4.14)$$

is convergent and

$$Mh + MK\frac{h^2}{2!} + \ldots + MK^{n-1}\frac{h^n}{n!} + \ldots \equiv \frac{M}{K}\left(e^{Kh}-1\right). \qquad (1.4.15)$$

As

- ♣ the terms of series (1.4.9) are majorized on I by positive constants, and
- ♣ the series of these constants is convergent,

the **Weierstrass criterion** can be applied (see e.g. [2][3][5][10]).

It follows that

The series (1.4.9) *is uniformly and absolutely convergent on* I.

Denote the sum of this series by y. Further on, taking into account the continuity of the general term of (1.4.9), by virtue of the properties of the sum of uniformly convergent series of functions (see e.g. [2][3][5][10]), it follows that

The sum *y* of the series (1.4.9) *is continuous*.

Consequently, one can pass to the limit in formula (1.4.5), thus getting

$$y(x) = y_0 + \int_{x_0}^{x} f(t, y(t)) \, dt; \qquad (1.4.16)$$

as *y* and *f* are continuous, it follows that the right side of (1.4.16) is differentiable, therefore the left side *y* is of class $C^1(I)$. Consequently, *y* satisfies the Cauchy problem (1.4.1).

* THE UNIQUENESS OF THE SOLUTION

Using a reductio ad absurdum argument, we assume that there exist two distinct solutions of the problem (1.4.1), let them be Y and Z. Then the modulus of their difference $\theta(x) \equiv |Y(x) - Z(x)|$ cannot vanish identically on the interval $[x_0, x_0 + \varepsilon]$. Therefore, the positive and continuous function θ cannot reach its maximum at a point $\xi \in [x_0, x_0 + \varepsilon]$. Yet ξ cannot coincide with x_0, because $\theta(x_0) = 0$. We have

$$\theta(\xi) \equiv |Y(\xi) - Z(\xi)|$$
$$\leq \int_{x_0}^{\xi} |f(t, Y(t)) - f(t, Z(t))| \, dt \qquad (1.4.17)$$
$$\leq K \int_{x_0}^{\xi} |Y(t) - Z(t)| \, dt = K \int_{x_0}^{\xi} \theta(t) \, dt \leq K\varepsilon\theta(\xi).$$

As $\theta(\xi) \neq 0$, (1.4.17) implies the absurd inequality $1 \leq K\varepsilon$, for ε arbitrarily small. ∎

* 1.4.2. THE CONTRACTION MAPPING PRINCIPLE

The successive approximation method, applied to ODEs, implies a much more general concept, with many applications: *the contraction mapping principle*. We shall briefly describe it.

Let X be a set on which a *distance* (*metric*) was defined:
$$d : X \times X \to \Re_+, \qquad (1.4.18)$$
with the properties:

1. $d(x, y) > 0$ și $d(x, y) = 0 \Leftrightarrow x = y$,

2. $d(x, y) = d(y, x), \forall x, y \in X$ – the property of *symmetry*,

3. $d(x, y) \leq d(x, z) + d(z, y), \forall x, y, z \in X$ – the *triangle inequality*.

Therefore, X along with d, which satisfies the above-mentioned properties, form *the metric space* (X, d).

Definitions:

1. The sequence $\{x_n\}_{n\in\mathfrak{N}} \subset X$ is convergent to $x \in X$ if the sequence of numbers $\{d(x_n, x)\}_{n\in\mathfrak{N}}$ converges to zero.

2. The sequence $\{x_n\}_{n\in\mathfrak{N}} \subset X$ is called a ***Cauchy sequence*** if, for any $\varepsilon > 0$, we can find an index $N(\varepsilon)$ such that

$$d(x_n, x_{n+p}) < \varepsilon, \qquad (1.4.19)$$

for any $n > N(\varepsilon)$ and any $p \in \mathfrak{N}$.

3. (X, d) is said to be a ***complete metric space*** if any Cauchy sequence allows a limit in X.

Let us now consider an operator $T : X \to X$.

Definitions:

1. T is called a ***contraction*** if there exists a positive number $\rho < 1$ such that

$$d(Tx, Ty) \leq \rho d(x, y), \qquad \forall x, y \in X. \qquad (1.4.20)$$

2. x is called a ***fixed point*** of T if

$$x = Tx. \qquad (1.4.21)$$

Taking into account these definitions and explanations, we can now prove

Theorem 1. 3. (***The contraction mapping principle, or the fixed point theorem***): *Let (X, d) be a complete metric space and consider the contraction $T : X \to X$. Then T admits a unique fixed point.*

Proof:

EXISTENCE: As T is a contraction, by virtue of the definition, it follows that

$$d(Tx, Ty) < \rho d(x, y), \quad \rho < 1. \tag{1.4.22}$$

Take $x_0 \in X$ arbitrary. We set up the sequence

$$\begin{aligned} x_1 &= Tx_0 \\ x_2 &= Tx_1 \\ &\ldots\ldots\ldots \\ x_n &= Tx_{n-1} \\ &\ldots\ldots\ldots \end{aligned} \tag{1.4.23}$$

Step by step, we get

$$\begin{aligned} d(x_1, x_0) &= d(Tx_0, x_0) = q \\ d(x_2, x_1) &= d(Tx_1, Tx_0) \le \rho d(x_1, x_0) = \rho q \\ d(x_3, x_2) &= d(Tx_2, Tx_1) \le \rho d(x_2, x_1) \le \rho^2 q \end{aligned} \tag{1.4.24}$$

$$\ldots\ldots\ldots\ldots\ldots\ldots\ldots\ldots\ldots\ldots\ldots\ldots\ldots\ldots$$

$$d(x_n, x_{n-1}) \le \rho^{n-1} q.$$

Then

$$\begin{aligned} d(x_n, x_{n+p}) &\le d(x_n, x_{n+1}) + d(x_{n+1}, x_{n+2}) + \ldots \\ &+ d(x_{n+p-1}, x_{n+p}) \le \rho^n q + \rho^{n+1} q + \ldots \\ &+ \rho^{n+p-1} q = \rho^n q (1 + \rho + \ldots + \rho^{p-1}), \end{aligned} \tag{1.4.25}$$

whence

$$d(x_n, x_{n+p}) \le q\rho^n \cdot \frac{1-\rho^p}{1-\rho} < q \frac{\rho^n}{1-\rho}, \quad p \in \mathfrak{N}. \tag{1.4.26}$$

As $\rho < 1$, for any $\varepsilon > 0$ arbitrarily chosen, but fixed up for the moment, there is an index $N(\varepsilon) \equiv N_\varepsilon$ such that

$$q \frac{\rho^{N_\varepsilon}}{1-\rho} < \varepsilon. \qquad (1.4.27)$$

Again, from $\rho < 1$, we obviously get

$$q \frac{\rho^n}{1-\rho} < \varepsilon, \quad n > N_\varepsilon. \qquad (1.4.28)$$

The inequality (1.4.26) satisfied by d leads to

$$d(x_n, x_{n+p}) < \varepsilon, \quad n > N_\varepsilon, p \in \mathfrak{N}. \qquad (1.4.29)$$

Therefore $\{x_n\}$ is a Cauchy sequence. But (X, d) is a complete metric space, so this sequence must allow a limit $x \in X$, i.e.,

$$\lim_{n \to \infty} d(x_n, x) = 0. \qquad (1.4.30)$$

This limit is a fixed point of T.

Indeed,

$$d(Tx_n, Tx) < \rho d(x_{n-1}, x) \implies \lim_{n \to \infty} d(Tx_n, Tx) = 0. \qquad (1.4.31)$$

From the definition $x_n = Tx_{n-1}$ it results that

$$d(x_n, x) \to 0, \quad d(Tx_{n-1}, Tx) \to 0, \qquad (1.4.32)$$

hence

$$\boxed{Tx = x}. \qquad (1.4.33)$$

UNIQUENESS: Let $x, y \in X$ be two distinct fixed points, hence $d(x, y) \neq 0$. If we simultaneously have

$$\begin{aligned} x &= Tx, \\ y &= Ty, \end{aligned} \quad (1.4.34)$$

then

$$d(x, y) = d(Tx, Ty) \leq \rho\, d(x, y), \quad (1.4.35)$$

whence we obtain

$$d(x, y) < d(x, y); \quad (1.4.36)$$

this relation is strict, fact which is absurd.

APPLICATION: PROOF OF THE CAUCHY-PICARD THEOREM USING THE CONTRACTION MAPPING PRINCIPLE

Theorem 1.4. *Consider the Cauchy problem*

$$\begin{cases} y' = f(x, y), \\ y(x_0) = y_0. \end{cases} \quad (1.4.37)$$

Suppose that the hypotheses of theorem 1.2 hold true, i.e.

1) $f \in C^0(D)$, $\quad D = \{(x, y) \,\|\, |x - x_0| < a,\ |y - y_0| < b\,\}$;

2) f is Lipschitz with respect to y on D, i.e., there exists a constant $K > 0$ such that

$$|f(x, Y) - f(x, Z)| < K|Y - Z|, \quad \forall (x, Y), (x, Z) \in D. \quad (1.4.38)$$

Then the Cauchy problem (1.4.37) locally allows a unique solution.

Proof: We integrate the equation from (1.4.37), as in the proof of theorem 1.2

$$\int_{x_0}^{x} y'(t) \, dt = \int_{x_0}^{x} f(t, y(t)) \, dt, \qquad (1.4.39)$$

thus getting

$$y(x) = y(x_0) + \int_{x_0}^{x} f(t, y(t)) \, dt, \qquad (1.4.40)$$

or, by virtue of the Cauchy condition,

$$y(x) = y_0 + \int_{x_0}^{x} f(t, y(t)) \, dt. \qquad (1.4.41)$$

The Cauchy problem (1.4.37) is thus equivalent to the integral equation (1.4.41). This emphasizes the operator

$$Ty \equiv y_0 + \int_{x_0}^{x} f(t, y(t)) \, dt, \qquad (1.4.42)$$

which is defined on the set (vector space) $C^0(I_0)$, its range being also in $C^0(I_0)$; by I_0 we mean the interval $I_0 = [x - a, x + a]$. Let

$$h = \min\left\{\frac{1}{K}, a\right\}, \qquad (1.4.43)$$

where K is the Lipschitz constant and consider the interval $I = [x_0 - h, x_0 + h]$.

Let us use the "sup" norm to define on $C^0(I_0)$ a metric by putting

$$d(x,y) = \sup_{x \in I} |y(x) - z(x)|, \quad y, z \in C^0(I). \tag{1.4.44}$$

Then

$$d(Ty, Tz) = \sup_{x \in I} \left| \int_{x_0}^{x} f(t, y(t)) dt - \int_{x_0}^{x} f(t, z(t)) dt \right|$$
$$\leq \sup_{x \in I} K \left| \int_{x_0}^{x} |y(t) - z(t)| dt \right| \leq K |x - x_0| \cdot d(y, z). \tag{1.4.45}$$

Hence

$$d(Ty, Tz) \leq Kh \, d(y, z), \tag{1.4.46}$$

where $Kh \equiv \rho < 1$, according to the inequality (1.4.43).

Therefore T is a **contraction**.

Let us notice that $C^0(I)$ is complete with respect to the above-defined metric, which is, in fact, with the "sup" norm.

We apply the contraction principle and it follows that there is a unique $Y \in C^0(I)$, so that

$$Y = TY, \tag{1.4.47}$$

i.e.

$$Y(x) = y_0 + \int_{x_0}^{x} f(t, Y(t)) dt. \tag{1.4.48}$$

But f is continuous, thus the right side of the primitive is of class $C^1(I)$. From this, it follows that $Y \in C^1(I)$, therefore Y satisfies (1.4.37). ∎

Example: Consider the Cauchy problem

$$\begin{cases} \dfrac{dy}{dx} = x^2 + y^2 \\ y(0) = 0 \end{cases}, \quad D = \{(x,y) \mid |x|<1, |y|<1\}. \qquad (1.4.49)$$

Approximate the solution of the Cauchy problem by using the method of successive approximations.

Solution:

STEP 1. We identify the data from the theorems 1.2 and 1.4:

$$f(x,y) = x^2 + y^2, \quad x_0 = 0, \quad y_0 = 0, \quad a = 1, \quad b = 1. \qquad (1.4.50)$$

STEP 2. We establish the interval on which the method is applicable.

a) According to the theorem 1.2, we have

$$h = \min\left\{a, \dfrac{b}{M}\right\}, \quad M = \sup_{(x,y)\in D}\{|f(x,y)|\}, \qquad (1.4.51)$$

where

$$M = \sup_{(x,y)\in D}\{|f(x,y)|\} = \sup_{|x|<1,|y|<1}\{|x^2+y^2|\} = 2, \qquad (1.4.52)$$

therefore

$$h = \min\left\{1, \dfrac{1}{2}\right\} = \dfrac{1}{2}. \qquad (1.4.53)$$

b) According to the theorem 1.4, we have

$$|f(x,y) - f(x,z)| = |x^2 + y^2 - x^2 - z^2| \le |y-z||y+z|, \qquad (1.4.54)$$

hence

$$|f(x,y)-f(x,z)| < 2|y-z|. \qquad (1.4.55)$$

Therefore $K = 2$ and, according to the inequality (1.4.43),

$$h = \min\left\{\frac{1}{K}, a\right\} = \min\left\{\frac{1}{2}, 1\right\} = \frac{1}{2}, \qquad (1.4.56)$$

i.e., the same value as in the case of theorem 1.2.

Thus, the required interval is

$$I \equiv \left[-\frac{1}{2}, \frac{1}{2}\right]. \qquad (1.4.57)$$

Let us compute the first three successive approximations of the solution of the problem (1.4.49). We have

$$y_0 = 0,$$
$$y_1 = 0 + \int_0^x (t^2 + 0^2) dt = \frac{x^3}{3},$$
$$y_2 = 0 + \int_0^x \left[t^2 + \left(\frac{t^3}{3}\right)^2\right] dt = \frac{x^3}{3} + \frac{x^7}{63},$$
$$y_3 = \int_0^x \left[t^2 + \left(\frac{t^3}{3} + \frac{t^7}{63}\right)^2\right] dt \qquad (1.4.58)$$
$$= \frac{x^3}{3} + \frac{x^7}{63} + \frac{2x^{11}}{2079} + \frac{x^{15}}{59535}.$$

We notice that the functions y_1, y_2, y_3 are odd and monotonically increasing. Therefore each one of them reaches its maximum at the point $x = \frac{1}{2}$. Computing the value of the approximants y_1, y_2, y_3 at this point, we get:

$$y_1\left(\frac{1}{2}\right) = \frac{1}{24} \cong 0,041666,$$

$$y_2\left(\frac{1}{2}\right) = 0,041666 + \frac{1}{63 \cdot 128} \cong 0,04179, \qquad (1.4.59)$$

$$y_3\left(\frac{1}{2}\right) = 0,04179 + \underbrace{\frac{1}{2^{10} \cdot 2079} + \frac{1}{2^{15} \cdot 59535}}_{<10^{-6}}.$$

Therefore, even for a small number of iterations (three), the approximating solutions are very close to each other !

Remarks.

- ♣ The values of h, which have been computed according to the two theorems 1.2 and 1.4, do not always coincide; this happens because we use the Lipschitz constant K on the one hand, and the maximum of the function f, on the other hand.

- ♣ Applying the method of successive approximation is similar to searching for the limit of a Cauchy sequence: we do not know the limit, but the further we advance into the sequence, the closer to each other are the terms, thus getting closer to the limit too.

Example. Take the simplest pocket computer with basic functions and press keys to write an arbitrary number on the display. Then press successively the "cos" key (compute in radians!). After several iterations, the display will change no more. **This means that you have solved the equation $x = \cos x$ with a precision of 10^{-7}!**

EXERCISES AND PROBLEMS

1. Find the general solution of the following ODE with separate variables:

a) $\dfrac{1}{1+x}dx + \dfrac{1}{1+y}dy = 0$ \qquad b) $(1+e^x)dx - \sin y\, dy = 0$

c) $\dfrac{x+1}{x}dx + \dfrac{y+1}{y}dy = 0$ \qquad d) $\dfrac{x}{1+x^2}dx - 2y\, dy = 0$

e) $\dfrac{dx}{x} + \dfrac{y}{2y-1}dy = 0$ \qquad f) $\sin x\, dx - \dfrac{2}{y^3}dy = 0$

2. Find the general solution, as well as the solution which satisfies the Cauchy condition (when required), of the following ODE with separable variables:

a) $(1+y^2)dx = x\, dy$ \qquad A: $y = \tan(\ln|x| + C)$

b) $y' = 2^{x-y}$ \qquad A: $2^x - 2^y = C$

c) $\dfrac{dx}{x(y-1)} + \dfrac{dy}{y(x+2)} = 0,\ y(1) = 1$ \qquad A: $x + y + 2\ln|x| - \ln|y| = 2$

d) $x\sqrt{1+y^2}dx + y\sqrt{1+x^2}dy = 0$ \qquad A: $\sqrt{1+x^2} + \sqrt{1+y^2} = C$

e) $\dfrac{x}{\sqrt{1-y^2}}dy + \dfrac{y}{\sqrt{1-x^2}}dx = 0$ \qquad A: $(1-x^2)^{\frac{3}{2}} + (1-y^2)^{\frac{3}{2}} = C$

f) $\dfrac{y}{x}y' + e^y = 0,\ y(1) = 0$ \hspace{2em} A: $2e^{-y}(y+1) = x^2 + 1$

g) $x(1+2y) + (1+x^2)y' = 0$ \hspace{2em} A: $(1+2y)(1+x^2) = C$

h) $\dfrac{y}{y'} = \ln y,\ y(2) = 1$ \hspace{2em} A: $2(x-2) = \ln^2 y$

i) $x(y^6+1)dx + y^2(x^4+1)dy = 0,\ y(0) = 1$

A:
$$3\arctan x^2 + 2\arctan y^3 = \dfrac{\pi}{2}$$

j) $\ln\cos y\, dx + x\tan y\, dy = 0$ \hspace{1em} A: $y = \arccos(e^{Cx})$

k) $\cos^2 y \cdot \tan y + \cos^2 x \cdot \tan x \dfrac{dy}{dx} = 0$ \hspace{1em} A: $\tan y \cdot \tan x = C$

l) $2yy' = \dfrac{e^x}{e^x + 1}$ \hspace{2em} A: $y^2 = \ln C(e^x + 1)$

m) $3e^x \tan y\, dx + (1+e^x)\sec^2 y\, dy = 0,\ y(0) = \dfrac{\pi}{4}$

Hint: $\sec^2\alpha = 1 + \tan^2\alpha$

A: $(1+e^x)^3 \cdot \tan y = 8$

n) $yy' = -2x\sec y$ \hspace{2em} A: $x^2 + y\sin y + \cos y = C$

o) $\sec^2 x \cdot \tan y\, dx + \sec^2 y \cdot \tan x\, dy = 0$, $y\left(\dfrac{\pi}{4}\right) = \dfrac{\pi}{4}$

$$A: \tan x \cdot \tan y = 1$$

p) $5e^x \tan y\, dx + (1 - e^x)\sec^2 y\, dy = 0$

$$A: y = \arctan C(1 - e^x)^5$$

r) $e^{1+x^2} \tan y\, dx - \dfrac{e^{2x}}{x-1}\, dy = 0$, $y(1) = \dfrac{\pi}{2}$

$$A: 2\ln|\sin y| = e^{(x-1)^2} - 1$$

s) $(1 + e^{2x})y^2\, dy = e^x\, dx$, $y(0) = 0$ $\quad A: \dfrac{1}{3}y^3 + \dfrac{\pi}{4} = \arctan e^x$

t) $y' + \cos(x + 2y) = \cos(x - 2y)$, $y(0) = \dfrac{\pi}{4}$

Hint: $\cos u - \cos v = -2\sin\dfrac{u+v}{2}\sin\dfrac{u-v}{2}$

$$A: \ln|\tan y| = 4(1 - \cos x)$$

u) $y' + \sin(x + y) = \sin(x - y)$

$$A: 2\sin x + \ln\left|\tan\dfrac{y}{2}\right| = C$$

Hint:

$\sin u \pm \sin v = 2\sin\dfrac{u \pm v}{2}\cos\dfrac{u \mp v}{2}$

3. Find the general solution, as well as the solution which satisfies the Cauchy condition (where required), of the following homogeneous ODE:

a) $x^2 y' = x^2 + xy - y^2$ \hspace{1cm} A: $x^2(y-x) = C(x+y)$

b) $2x^3 y' = 3x^2 y + y^3$, $y(1) = 1$ \hspace{1cm} A: $y^2 = \frac{1}{2}|x|(y^2 + x^2)$

c) $y' = \dfrac{x^3 - 3xy^2}{3x^2 y - y^3}$ \hspace{1cm} A: $y^4 - 6y^2 x^2 + x^4 = C$

d) $xyy' = y^2 + 2x^2$ \hspace{1cm} A: $y^2 = 2x^2 \ln(Cx^2)$

e) $(x^2 + y^2)dx - xy\,dy = 0$ \hspace{1cm} A: $y^2 = x^2 \ln(Cx^2)$

f) $2x^2 y' = x^2 + y^2$ \hspace{1cm} A: $y = x - \dfrac{2x}{C + \ln|x|}$

g) $y' = \dfrac{x+y}{x-y}$ \hspace{1cm} A: $\arctan\dfrac{y}{x} = \ln\left(C\sqrt{x^2 + y^2}\right)$

h) $\left(x^4 + 6x^2 y^2 + y^4\right)dx + 4xy\left(x^2 + y^2\right)dy = 0$, $y(1) = 0$

\hspace{4cm} A: $x^5 + 10x^3 y^2 + 5xy^4 = 1$

i) $xy' = 2\left(y - \sqrt{xy}\right)$ \hspace{1cm} A: $16xy = \left(y + 4x - Cx^2\right)^2$

j) $x\sin\dfrac{y}{x} \cdot y' + x = y\sin\dfrac{y}{x}$ \hspace{1cm} A: $Cx = e^{\cos\frac{y}{x}}$

k) $xy' \ln\dfrac{y}{x} = x + y\ln\dfrac{y}{x}$ \hspace{1cm} A: $\ln|x| = \dfrac{y}{x}\left(\ln\left|\dfrac{y}{x}\right| - 1\right) + C$

l) $y' = \dfrac{y}{x} + \cos\dfrac{y}{x}$ A: $\dfrac{1+\sin\dfrac{y}{x}}{1-\sin\dfrac{y}{x}} = Cx^2$

m) $xy' = y - xe^{\frac{y}{x}}$ A: $y = -x\ln|C + \ln|x||$

n) $3y\sin\dfrac{3x}{y}dx + \left(y - 3x\sin\dfrac{3x}{y}\right)dy = 0$

A: $\ln|y| - \cos\dfrac{3x}{y} = C$

4. Find the general solution, as well as the solution which satisfies the Cauchy condition (where required), of the following exact/total differential equations:

a) $(x+y)dx + (x-y)dy = 0$ A: $\dfrac{x^2}{2} + xy - \dfrac{y^2}{2} = C$

b) $(3x^2 + 6xy^2)dx + (6x^2y + 4y^3)dy = 0$

A: $x^3 + 3x^2y^2 + y^4 = C$

c) $8xy - 5y^2 + 2x(2x - 5y)y' = 0$ A: $4x^2y - 5xy^2 = C$

d) $3x(x + 2y^2)dx + 2y(3x^2 + 2y^2)dy = 0$

A: $x^3 + 3x^2y^2 + y^4 = C$

e) $ye^x dx + (y + e^x)dy = 0$

A: $ye^x + \dfrac{y^2}{2} = C$

f) $(e^x + y + \sin y)dx + (e^y + x + x\cos y)dy = 0$

A: $e^x + xy + x\sin y + e^y = C$

g) $(x+y-1)dx + (e^y + x)dy = 0$

$$A: \frac{1}{2}x^2 + xy - x + e^y = C$$

h) $(y + e^x \sin y)dx + (x + e^x \cos y)dy = 0$

$$A: xy + e^x \sin y = C$$

i) $(e^x \sin y + x)dx + (e^x \cos y + y)dy = 0$

$$A: x^2 + y^2 + 2e^x \sin y = C$$

j) $(x^2 + y^2 + y)dx + (2xy + x + e^y)dy = 0,\ y(0) = 0$

$$A: \frac{1}{3}x^3 + xy^2 + xy + e^y = 1$$

k) $\left(\dfrac{y}{x^2+y^2} - y\right)dx + \left(e^y - x - \dfrac{x}{x^2+y^2}\right)dy = 0$

$$A: \arctan\frac{x}{y} - xy + e^y = C$$

l) $(e^{x+y} + 3x^2)dx + (e^{x+y} + 4y^3)dy = 0,\ y(0) = 0$

$$A: e^{x+y} + x^3 + y^4 = 1$$

m) $(2xye^{x^2} + \ln y)dx + \left(e^{x^2} + \dfrac{x}{y}\right)dy = 0,\ y(0) = 1$

$$A: ye^{x^2} + x \ln y = 1$$

n) $\left(1+e^{\frac{x}{y}}\right)dx + e^{\frac{x}{y}}\left(1-\frac{x}{y}\right)dy = 0$

$$A: x + ye^{\frac{x}{y}} = C$$

o) $(x^2 + \sin y)dx + (1 + x\cos y)dy = 0$

$$A: x^3 + 3y + 3x\sin y = C$$

p) $[\sin y + (1-y)\cos x]dx + [(1+x)\cos y - \sin x]dy = 0$

$$A: (1+x)\sin y + (1-y)\sin x = C$$

r) $(y + x\ln y)dx + \left(\dfrac{x^2}{2y} + x + 1\right)dy = 0$

$$A: x^2 \ln y + 2y(x+1) = C$$

s) $(\ln y - 5y^2 \sin 5x)dx + \left(\dfrac{x}{y} + 2y\cos 5x\right)dy = 0,\ y(0) = e$

$$A: x\ln y + y^2 \cos 5x = e^2$$

t) $(3x^2 y + \sin x)dx + (x^3 - \cos y)dy = C$

$$A: x^3 y - \cos x - \sin y = C$$

u) $(xy + \sin y)dx + \left(\dfrac{1}{2}x^2 + x\cos y\right)dy = 0$

$$A: \dfrac{1}{2}x^2 y + x\sin y = C$$

v) $(x+\sin y)dx+(x\cos y+\sin y)dy=0$

$$A: \frac{1}{2}x^2 + x\sin y - \cos y = C$$

5. Solve the following ODE, searching for an integrating factor of the form $\mu = \mu(x,y)$:

a) $(x^2+y)dx - xdy = 0, \mu = \mu(x)$ $A: x - \dfrac{x}{y} = C$, $\mu = \dfrac{1}{x^2}$

b) $(2x^3y^2 - y)dx + (2x^2y^3 - x)dy = 0$,

$\mu = \mu(xy)$

$$A: x^2 + y^2 + \frac{1}{xy} = C, \ \mu = \frac{1}{x^2y^2}$$

c) $(x\sin y + y\cos y)dx + (x\cos y - y\sin y)dy = 0$,

$\mu = \mu(x)$

$$A: e^x(x\sin y + y\cos y - \sin y) = C,$$

$$\mu = e^x$$

d) $ydx - xdy + \ln x\, dx = 0$,

$\mu = \mu(x)$

$$A: y = Cx - \ln x - 1, \ \mu = \frac{1}{x^2}$$

e) $(x^2\cos x - y)dx + xdy = 0$,

$\mu = \mu(x)$

$$A: y = x(C - \sin x), \ \mu = \frac{1}{x^2}$$

f) $y\,dx - (x + y^2)\,dy = 0, \mu = \mu(y)$ \qquad A: $x = y(C + y)$,

$$\mu = \frac{1}{y^2}$$

g) $y\sqrt{1-y^2}\,dx + \left(x\sqrt{1-y^2} + y\right)dy = 0$, \qquad A: $xy - \sqrt{1-y^2} = C$,

$\mu = \mu(y)$ \qquad\qquad $\mu = \dfrac{1}{\sqrt{1-y^2}}$

6. Find the general solution, as well as the solution which satisfies the Cauchy condition (when required), for the following first order linear ODEs:

a) $xy' + y = x + 1$ \qquad A: $y = \dfrac{C}{x} + \dfrac{x}{2} + 1$

b) $xy' - y + x\sin x + \cos x = 0$ \qquad A: $y = Cx + \cos x$

c) $y' + y\cos x = \sin x \cos x$ \qquad A: $y = Ce^{-\sin x} + \sin x - 1$

d) $y' + y\sin x = -\sin x \cos x$ \qquad A: $y = -\cos x - 1 + Ce^{\cos x}$

e) $xy' - y = x^2 \cos x$ \qquad A: $y = x(\sin x + C)$

f) $y' + 2xy = xe^{-x^2}$ \qquad A: $y = e^{-x^2}\left(\dfrac{1}{2}x^2 + C\right)$

g) $y'\cos x + y = 1 - \sin x$ \qquad A: $y = \dfrac{(\sin x + C)\cos x}{1 + \sin x}$

h) $y' + \dfrac{n}{x}y = \dfrac{a}{x^n}$, $y(1) = 0$ \qquad A: $y = \dfrac{a(x-1)}{x^n}$

i) $(1 + x^2)y' + y = \arctan x$ \qquad A:

$$y = \arctan x - 1 + Ce^{-\arctan x}$$

j) $y'\sqrt{1-x^2} + y = \arcsin x,$
$y(0) = 0$

A: $y = e^{-\arcsin x} + \arcsin x - 1$

k) $y' - \dfrac{y}{\sin x} = \cos^2 x \tan \dfrac{x}{2}$

A: $y = \tan \dfrac{x}{2}\left(\dfrac{x}{2} + \dfrac{1}{4}\sin 2x + C\right)$

l) $y' - \dfrac{y}{x \ln x} = x \ln x,$
$y(e) = \dfrac{1}{2}e^2$

A: $y = \dfrac{1}{2}x^2 \ln x$

m) $y' \sin x - y \cos x = 1,$
$y\left(\dfrac{\pi}{2}\right) = 0$

A: $y = -\cos x$

7. Find the general solution, as well as the solution which satisfies the Cauchy condition (when required), for the following Bernoulli equations:

a) $x y' + y = y^2 \ln x$

A: $y = \dfrac{1}{Cx + \ln x + 1}$

b) $4xy' + 3y = -e^x x^4 y^5$

A: $y^{-4} = x^3\left(e^x + C\right)$

c) $x y' - 4y = 2x^3 \sqrt{y}$

A: $y = \left(Cx^2 + x^3\right)^2$

d) $y' + \dfrac{2y}{x} = 3x^2 y^{\tfrac{4}{3}}$

A: $y^{-\tfrac{1}{3}} = Cx^{\tfrac{2}{3}} - \dfrac{3}{7}x^3$

e) $y' - \dfrac{y}{x-1} = \dfrac{y^2}{x-1}$

A: $y = \dfrac{x-1}{C-x}$

f) $y' + \dfrac{2y}{x} = \dfrac{2\sqrt{y}}{\cos^2 x}$ $\qquad$ A: $\sqrt{y} - \tan x = \dfrac{\ln\cos + C}{x}$

g) $y' + y = e^{\frac{1}{2}x}\sqrt{y}$, $y(0) = \dfrac{9}{4}$ $\qquad$ A: $y = e^{-x}\left(\dfrac{1}{2}e^x + 1\right)^2$

h) $y' + \dfrac{3x^2 y}{x^3 + 1} = y^2(x^3 + 1)\sin x$, $\qquad$ A: $y = \dfrac{\sec x}{x^3 + 1}$
$y(0) = 1$.

i) $y' - 2y\tan x + y^2 \sin^2 x = 0$ $\qquad$ A: $y = \dfrac{\sec^2 x}{\tan x - x + C}$

8. Find the general solution of the following Riccati equations, which allow the particular solution $Y = -\dfrac{1}{x}$:

a) $x^2 y' = x^2 y^2 + xy + 1$ $\qquad$ A: $y = -\dfrac{1}{x} + \dfrac{1}{Cx - x\ln x}$

b) $y' + y^2 = \dfrac{2}{x^2}$ $\qquad$ A: $y = \dfrac{2x^3 - 3C}{x(3C + x^3)}$

9. Integrate the following Clairaut's equations:

$\qquad\qquad\qquad\qquad$ A: gen. sol.: $y = Cx - e^C$

a) $y = xy' - e^{y'}$ $\qquad$ sing.sol.: $\begin{cases} x = e^p \\ y = (p-1)e^p \end{cases} \leftrightarrow y = x(\ln x - 1)$

$\qquad\qquad\qquad\qquad$ A: gen. sol.: $y = xC + C - C^2$

b) $y = xy' + y' - y'^2$ $\qquad$ sing.sol.: $\begin{cases} x = 2p - 1 \\ y = p^2 \end{cases} \leftrightarrow y = \dfrac{1}{4}(x+1)^2$

A: gen. sol.: $y = xC - \dfrac{1}{C}$

c) $x = \dfrac{y}{y'} + \dfrac{1}{y'^2}$

sing.sol.: $\begin{cases} x = -\dfrac{1}{p^2} \\ y = -\dfrac{2}{p} \end{cases} \leftrightarrow y^2 = -4x$

A: gen. sol.: $y = Cx + C^2$

d) $y = xy' + y'^2$

sing.sol.: $\begin{cases} x = -2p \\ y = -p^2 \end{cases} \leftrightarrow y = -\dfrac{x^2}{4}$

A: gen. sol.: $y = Cx + \dfrac{1}{C^2}$

e) $y = xy' + \dfrac{1}{y'^2}$

sing.sol.: $\begin{cases} x = \dfrac{2}{p^3} \\ y = \dfrac{3}{p^2} \end{cases} \leftrightarrow 4y^3 = 27x^2$

A: gen. sol.: $y = Cx + C^2 + 1$

f) $y = x\left(\dfrac{1}{x} + y'\right) + y'^2$

sing.sol.: $\begin{cases} x = -2p \\ y = 1 - p^2 \end{cases} \leftrightarrow y = 1 - \dfrac{x^4}{4}$

A: gen. sol.: $y = Cx + \sqrt{b^2 + a^2 C^2}$

g) $y = xy' + \sqrt{b^2 + a^2 y'^2}$ sing.sol.: $\begin{cases} x = -\dfrac{a^2 p}{\sqrt{b^2 + a^2 p^2}} \\ y = \dfrac{b^2}{\sqrt{b^2 + a^2 p^2}} \end{cases}$

$\leftrightarrow \dfrac{x^2}{a^2} + \dfrac{y^2}{b^2} = 1$

10. Integrate the following Lagrange's equations:

a) $y = xy'^2 + y'^2$

A: gen. sol.: $\begin{cases} x+1 = \dfrac{C}{(p-1)^2} \\ y = \dfrac{Cp^2}{(p-1)^2} \end{cases}$ $\leftrightarrow \left(\sqrt{y} + \sqrt{x+1}\right)^2 = C$

sing.sol.: $y = 0$

b) $2y(y'+1) = xy'^2$

A: gen. sol.: $\begin{cases} x = C(p+1) \\ y = \dfrac{C}{2}p^2 \end{cases}$ $\leftrightarrow y = \dfrac{(x-C)^2}{2C}$

sing.sol.: $y = 0$, $y = -2x$

Chapter 2

LINEAR ORDINARY DIFFERENTIAL EQUATIONS OF ORDER n

2.1. PRELIMINARY CONCEPTS. EXAMPLES

The general form of a linear ODE of order n is

$$Ly \equiv a_0(x)y^{(n)} + a_1(x)y^{(n-1)} + \ldots + a_{n-1}(x)y' + a_n(x)y = F(x) \quad (2.1.1)$$

where

$$a_j \in C^0(I), \; j = \overline{0,n}, \qquad F \in C^0(I), \quad I \subseteq \Re. \quad (2.1.2)$$

If $a_0(x) \neq 0$, $x \in I$, we divide by a_0 and we get

$$Ly \equiv y^{(n)} + p_1(x)y^{(n-1)} + \ldots + p_{n-1}(x)y' + p_n(x)y \\ = f(x) \quad (2.1.3)$$

where the the following notations were used:

$$p_j(x) = \frac{a_j(x)}{a_0(x)}, \; j = \overline{1,n}, \quad f(x) = \frac{F(x)}{a_0(x)}. \quad (2.1.4)$$

Suppose that there are some points x at which $a_0(x) = 0$. At these points, the equation "loses" its order; they are called **singular points**.

The equations with singular points do not form the object of this book. Therefore, we shall only use the form (2.1.3) of the linear ODE.

We **remind** that an operator $L: X \to Y$, where X, Y are vector subspaces, is called ***linear*** if

$$L(\alpha x_1 + \beta x_2) = \alpha L x_1 + \beta L x_2, \forall \alpha, \beta \in \mathfrak{R}/\mathbb{C}, \forall x_1, x_2 \in X. \quad (2.1.5)$$

Let us **prove** that the operator L defined by (2.1.1) is a ***linear operator***.

Indeed, let us take $y, z \in C^n(I)$, $\alpha, \beta \in \mathfrak{R}/\mathbb{C}$. We get

$$L(\alpha y + \beta z) = (\alpha y + \beta z)^{(n)} + p_1(x)(\alpha y + \beta z)^{(n-1)} + \ldots +$$
$$+ p_{n-1}(x)(\alpha y + \beta z)' + p_n(x)(\alpha y + \beta z) =$$
$$= \alpha y^{(n)} + \beta z^{(n)} + p_1(x)\left(\alpha y^{(n-1)} + \beta z^{(n-1)}\right) + \ldots +$$
$$+ p_{n-1}(x)(\alpha y' + \beta z') + p_n(x)(\alpha y + \beta z) =$$
$$= \alpha \left(\underbrace{y^{(n)} + p_1(x) y^{(n-1)} + \ldots + p_{n-1}(x) y' + p_n(x) y}_{Ly} \right) + \quad (2.1.6)$$
$$+ \beta \left(\underbrace{z^n + p_1(x) z^{(n-1)} + \ldots + p_{n-1}(x) z' + p_n(x) z}_{Lz} \right),$$

i.e.,

$$L(\alpha y + \beta z) = \alpha L y + \beta L z, \quad (2.1.7)$$

which is exactly what we had to prove.

As in the case of first order ODE, **we identify a linear operator by the fact that the unknown function and its derivatives, up to the nth order included, are of degree** 1.

Therefore, an ordinary differential equation of order $n \geq 2$ is linear if it is defined by a linear differential operator.

Examples

♣ The ODE $y''' + yy' = \sin x$ is **nonlinear**, due to the term yy', which is a second-degree monomial in y and y'.

♣ The ODE $y^{(4)} + x^2 y' + x^4 e^x y = 0$ is **linear**, because it is of first degree with respect to y, y' and $y^{(4)}$.

The equations (2.1.1), (2.1.3) are **linear**, because the differential operator $L : C^n(I) \to C^0(I)$ is linear; it is seen that in these equations the unknown function y and its derivatives up to the order n are of degree 1.

GENERAL PROPERTIES OF LINEAR ODEs OF ORDER n

1. *Any nonsingular change of variable transforms a linear ODE into another linear ODE of the same order.*

* Indeed, consider the change of variable

$$x = f(t), \quad f \in C^n([\alpha, \beta]), \quad [\alpha, \beta] \subseteq \Re, \qquad (2.1.8)$$

with $f'(t) \neq 0, t \in [\alpha, \beta]$. By virtue of the implicit function theorem (see e.g. [2][3][5][10]), the inverse transformation $t = \varphi(x)$ exists.

Let us compute the successive derivatives of y with respect to the new variable t. We get

$$\frac{dy}{dx} = \frac{dy}{dt}\frac{dt}{dx} = \frac{1}{f'(t)}\frac{dy}{dt},$$

$$\frac{d^2y}{dx^2} = \frac{1}{f'(t)}\frac{d}{dt}\left[\frac{1}{f'(t)}\frac{dy}{dt}\right] = \frac{1}{f'^2(t)}\frac{d^2y}{dt^2} - \frac{f''(t)}{f'^3(t)}\frac{dy}{dt};$$

(2.1.9)

therefore, the derivatives with respect to x are **linear** with respect to the derivatives in t.

Further differentiation also results in linear expressions, which, introduced in (2.1.1), finally lead to a linear ODE of the same order.

2. Any linear change of function in a linear ODE preserves its linearity and order.

For easier computation, let us take the second order linear equation

$$Ly \equiv y'' + p_1(x)y' + p_2(x)y = f(x). \qquad (2.1.10)$$

Consider the change

$$y = q(x)z(x) + r(x), \quad q, r \in C''([a,b]). \qquad (2.1.11)$$

Differentiating this twice, we obtain

$$\begin{array}{l|l}
y = q(x)z(x) + r(x) & \times p_2(x) \\
y' = qz' + q'z + r' & \times p_1(x) \\
y'' = qz'' + 2q'z' + q''z + r'' & \times 1 \\
\hline
Ly = qz'' + (qp_1 + 2q')z' + zLq + Lr.
\end{array} \qquad (2.1.12)$$

From the last expression, we get a differential equation of unknown function z

$$qz'' + (qp_1 + 2q')z' + zLq = -Lr + f, \qquad (2.1.13)$$

which is a second order linear ODE.

This result can also be proven in the same way for a linear ODE of an arbitrary order n.

2.2. LINEAR HOMOGENEOUS ODES OF ORDER n

The equations (2.1.1) and (2.1.3) are **non-homogeneous**, as they both have nonzero free terms.

We can associate them the corresponding homogeneous equations, by setting the right sides to zero, as follows:

The homogeneous ODE

$$Ly \equiv a_0(x)y^{(n)} + a_1(x)y^{(n-1)} + \ldots + a_{n-1}(x)y' + a_n(x)y = 0, \qquad (2.2.1)$$

corresponds to the equation (2.1.1); we associate *the homogeneous ODE*

$$Ly \equiv y^{(n)} + p_1(x)y^{(n-1)} + \ldots + p_{n-1}(x)y' + p_n(x)y = 0 \qquad (2.2.2)$$

to the equation (2.1.3).

As mentioned before, from now on we shall deal with the equation (2.2.2).

The kernel of the operator L is defined as

$$\ker L = \{y \in C^n(I) | L y = 0\} \subset C^n(I).$$

In other words, $\ker L$ is the set of solutions of the nth order homogeneous linear equation (2.2.2).

Theorem 2.1. $\ker L$ *is a linear subspace of* $C^n(I)$.

Proof. Let us take $y, z \in \ker L$. This means that $L y = 0$, $L z = 0$ on I.

But L is linear, therefore

$$L(\alpha y + \beta z) = \alpha \underbrace{L y}_{=0} + \beta \underbrace{L z}_{=0} = 0, \qquad (2.2.3)$$

whence it follows that $(\alpha y + \beta z) \in \ker L$. ∎

According to the properties of vector spaces, we can state that:

- As $\ker L$ is a vector subspace, any element from $\ker L$ is expressed as a linear combination of elements of a basis of $\ker L$.
- In order to solve the homogeneous ODE (2.2.2), we only need to find a **basis** in $\ker L$.

We can prove that the dimension of $\ker L$ is n, i.e.

$$\boxed{\dim \ker L = n}. \qquad (2.2.4)$$

This fact has also an obvious intuitive explanation. If, by integration, the first order derivative introduces an arbitrary constant, then the nth order derivative introduces n arbitrary constants (meaning n degrees of freedom).

Therefore, a basis in $\ker L$ is formed by n linearly independent functions of $\ker L$, or, equivalently, by n linearly independent solutions of the homogeneous ODE (2.2.2).

Definition 2.1. *A basis in* $\ker L$ *is called a fundamental system of solutions of the equation* (2.2.2).

We **remind** the definition of the *linear independence* of a system of functions.

Consider the system of functions $\{y_j\}_{j=\overline{1,n}} \subset C^0(I)$.

Definition 2.2. The system $\{y_j\}_{j=\overline{1,n}}$ is called *linearly dependent* if there exist some real constants $c_1, c_2, \ldots, c_n$, not all of them null (specifically, $\sum_{j=1}^{n} c_j^2 \neq 0$), such that

$$c_1 y_1(x) + c_2 y_2(x) + \ldots + c_n y_n(x) = 0, \quad \forall x \in I. \qquad (2.2.5)$$

Otherwise, the system is called *linearly independent*, i.e.

Definition 2.3. The system $\{y_j\}_{j=\overline{1,n}}$ is called *linearly independent* if the equality

$$c_1 y_1(x) + c_2 y_2(x) + \ldots + c_n y_n(x) = 0, \quad \forall x \in I, \qquad (2.2.6)$$

implies

$$c_j = 0, \quad j = \overline{1,n}. \qquad (2.2.7)$$

Examples

1. Prove that the functions $y_1 = 1$, $y_2 = \cos^2 x$, $y_3 = \sin^2 x$ form a *linearly dependent system* on $\Re$.

Indeed, by using the well-known trigonometric formula $\cos^2 x + \sin^2 x = 1$, it follows that the linear combination that proves the linear dependence of the above-mentioned system is:

$$y_2 + y_3 - y_1 = 0, \forall x \in \Re. \qquad (2.2.8)$$

2. Prove that the system of functions $\{1, x, x^2, x^3\}$ form a *linearly independent system* on $\Re$.

Indeed, if

$$c_1 \cdot 1 + c_2 \cdot x + c_3 \cdot x^2 + c_4 \cdot x^3 = 0, \quad \forall x \in \Re, \qquad (2.2.9)$$

then the left side of (2.2.9) is the zero polynomial, therefore its coefficients are all of them null:

$$c_j = 0, \quad j = \overline{1,4}. \qquad (2.2.10)$$

HOW DO WE VERIFY THE LINEAR INDEPENDENCE/DEPENDENCE OF A SYSTEM OF FUNCTIONS ?

For a better understanding, let us take $n = 3$; the case of arbitrary n can be treated analogously.

Therefore, we consider the system $\{y_1, y_2, y_3\} \in C^3(I)$. If

$$c_1 y_1(x) + c_2 y_2(x) + c_3 y_3(x) = 0, \qquad (2.2.11)$$

then its derivatives are also null:

$$\begin{aligned} c_1 y_1'(x) + c_2 y_2'(x) + c_3 y_3'(x) &= 0, \\ c_1 y_1''(x) + c_2 y_2''(x) + c_3 y_3''(x) &= 0, \end{aligned} \qquad (2.2.12)$$

for any $x \in I$.

The three relations (2.2.11), (2.2.12) form a homogeneous linear algebraic system, with c_1, c_2, c_3 as unknowns.

The associated determinant is

$$W[y_1, y_2, y_3] \equiv \begin{vmatrix} y_1(x) & y_2(x) & y_3(x) \\ y_1'(x) & y_2'(x) & y_3'(x) \\ y_1''(x) & y_2''(x) & y_3''(x) \end{vmatrix}, \quad x \in I, \quad (2.2.13)$$

and it is called *the Wronskian*.

According to algebra for linear systems, it follows that

- If $W \equiv 0$ on I, then the above-mentioned linear algebraic system allows non-zero solutions, therefore $\{y_1, y_2, y_3\}$ form a linearly dependent system;
- If $W \neq 0$ on I, then the system allows only the identically zero solution, therefore $\{y_1, y_2, y_3\}$ form a linearly independent system.

Let $\{y_1, y_2, y_3\}$ be solutions of the linear ODE

$$Ly \equiv y''' + p_1(x)y'' + p_2(x)y' + p_3(x)y = 0. \quad (2.2.14)$$

We can prove that:

Theorem 2.2. *If the system* $\{y_1, y_2, y_3\} \subset \ker L$ *is linearly independent, then the Wronskian is identically null on* I, *i.e.,* $W[y_1, y_2, y_3] \neq 0, \ \forall x \in I.$

* **Proof.** Take $x_0 \in I$. Using a reductio ad absurdum argument, we assume that $W(x_0) = 0$. Then there exist c_1, c_2, c_3, not all of them null, satisfying

$$\begin{aligned} c_1 y_1(x_0) + c_2 y_2(x_0) + c_3 y_3(x_0) &= 0, \\ c_1 y_1'(x_0) + c_2 y_2'(x_0) + c_3 y_3'(x_0) &= 0, \\ c_1 y_1''(x_0) + c_2 y_2''(x_0) + c_3 y_3''(x_0) &= 0. \end{aligned} \quad (2.2.15)$$

Consider the function $\tilde{y} = c_1 y_1(x) + c_2 y_2(x) + c_3 y_3(x)$. Then $\tilde{y}$ satisfies

$$\begin{cases} L\tilde{y} = 0, \\ \tilde{y}(x_0) = 0, \; \tilde{y}'(x_0) = 0, \; \tilde{y}''(x_0) = 0, \end{cases} \quad (2.2.16)$$

on I. Consequently, as the solution of this problem is unique and as the identically zero function is a solution, it results that $\tilde{y}(x) = 0$, $x \in I$, i.e.,

$$c_1 y_1(x) + c_2 y_2(x) + c_3 y_3(x) = 0, \quad x \in I, \quad (2.2.17)$$

with c_1, c_2, c_3 not all of them null. It follows that $\{y_1, y_2, y_3\}$ ***is not linearly independent***, which is a ***contradiction***!

Therefore $W(x) \neq 0, \forall x \in I$. ∎

Consider now the system of functions $\{y_j\}_{j=\overline{1,n}}$, at least of $C^{n-1}(I)$ class.

Definition 2.4. The determinant

$$W[y_1, y_2, \ldots, y_n] \equiv \begin{vmatrix} y_1(x) & y_2(x) & \ldots & y_n(x) \\ y_1'(x) & y_2'(x) & \ldots & y_n'(x) \\ \ldots & \ldots & \ldots & \ldots \\ y_1^{(n-1)}(x) & y_2^{(n-1)}(x) & \ldots & y_n^{(n-1)}(x) \end{vmatrix} \quad (2.2.18)$$

is called *the Wronskian* of the functions $\{y_1, y_2, \ldots, y_n\}$.

The theorem 2.2, as well as the above-mentioned statements regarding the Wronskian of a system of three functions, can easily be proven even for an arbitrary n.

In conclusion, for a system $\{y_1, y_2, \ldots, y_n\} \subset \ker L$, with L given by (2.2.2), the following alternative is valid

ALTERNATIVE:

- either $W[y_1, y_2, \ldots, y_n] \equiv 0$ on I and then the system $\{y_1, y_2, \ldots, y_n\}$ is linearly dependent;
- or $W[y_1, y_2, \ldots, y_n] \neq 0, x \in I$ and then the system $\{y_1, y_2, \ldots, y_n\}$ is linearly independent.

Examples

1. Let un take the equation

$$Ly \equiv y'' - y = 0 \quad (2.2.19)$$

and consider its system of solutions $y_1 = e^x$, $y_2 = e^{-x}$.

VERIFICATION. Indeed, we have

$$\begin{aligned} Ly_1 &= Le^x = e^x - e^x = 0, \quad \forall x \in \Re, \\ Ly_2 &= Le^{-x} = -(-e^{-x}) - e^{-x} = 0, \quad \forall x \in \Re. \end{aligned} \quad (2.2.20)$$

By definition, their Wronskian is

$$W\left[e^x, e^{-x}\right] = \begin{vmatrix} y_1 & y_2 \\ y_1' & y_2' \end{vmatrix} = \begin{vmatrix} e^x & e^{-x} \\ e^x & -e^{-x} \end{vmatrix} = -2 \neq 0. \qquad (2.2.21)$$

Therefore, according to the above-mentioned alternative, the system $\{e^x, e^{-x}\}$ is *a fundamental system* for the ODE (2.2.19), or, equivalently, *a basis* in $\ker L$. This means that the general solution of (2.2.19) is $\boxed{y(x) = c_1 e^x + c_2 e^{-x}}$, with c_1, c_2 arbitrary constants.

2. Take the equation

$$Ly \equiv y'' + y = 0. \qquad (2.2.22)$$

The functions $y_1 = \sin x$, $y_2 = \cos x$ form a system of solutions for this equation.

VERIFICATION. We have

$$\begin{aligned} Ly_1 &= (\cos x)' + \sin x = 0, \\ Ly_2 &= (-\sin x)' + \cos x = 0, \end{aligned} \qquad (2.2.23)$$

whence $\{y_1, y_2\} \subset \ker L$.

Now, we compute the Wronskian

$$W[\sin x, \cos x] = \begin{vmatrix} \sin x & \cos x \\ \cos x & -\sin x \end{vmatrix} = -1 \neq 0, \qquad (2.2.24)$$

which means that $\{\sin x, \cos x\}$ form *a basis* in $\ker L$ or, in other words, *a fundamental system of solutions*. This means that the general solution of (2.2.22) is $\boxed{y(x) = c_1 \cos x + c_2 \sin x}$, with c_1, c_2 arbitrary constants.

Let $\{y_j\}_{j=\overline{1,n}} \subset \ker L$, with given by (2.2.2), be a basis in $\ker L$. Then any solution y of the linear homogeneous ODE (2.2.2) is expressed as a linear combination of $y_j(x)$, i.e.

$$y(x) = c_1 y_1(x) + c_2 y_2(x) + \ldots + c_n y_n(x),$$
$$x \in I, c_j \in \mathfrak{R}, j = \overline{1,n}. \quad (2.2.25)$$

Therefore, we can conclude that

The general solution of the homogeneous ODE

$$Ly \equiv y^{(n)} + p_1(x) y^{(n-1)} + \ldots + p_{n-1}(x) y' +$$
$$+ p_n(x) y = 0, \qquad x \in I, \quad (2.2.26)$$

is of the form

$$y(x) = c_1 y_1(x) + c_2 y_2(x) + \ldots + c_n y_n(x), \quad (2.2.27)$$

where c_j **are arbitrary constants and** $\{y_j\}_{j=\overline{1,n}}$ **form a fundamental system of solutions of the ODE.**

Remark. Consider $y_1 = \operatorname{ch} x$, $y_2 = \operatorname{sh} x$ and the ordinary differential equation

$$Ly \equiv y'' - y = 0,$$

from the previous example.

We have

$$Ly_1 = (\operatorname{ch} x)'' - \operatorname{ch} x = \operatorname{ch} x - \operatorname{ch} x = 0,$$
$$Ly_2 = (\operatorname{sh} x)'' - \operatorname{sh} x = \operatorname{sh} x - \operatorname{sh} x = 0,$$

therefore $\{y_1, y_2\} \subset \ker L$.

The Wronskian of the system $\{y_1, y_2\}$ is

$$W[y_1, y_2] = \begin{vmatrix} \operatorname{ch} x & \operatorname{sh} x \\ \operatorname{sh} x & \operatorname{ch} x \end{vmatrix} = \operatorname{ch}^2 x - \operatorname{sh}^2 x = 1 \neq 0,$$

therefore $\{\operatorname{ch} x, \operatorname{sh} x\}$ form *a fundamental system of solutions* of (2.2.19). But we have shown that $\{e^x, e^{-x}\}$ also form a fundamental system of solutions for the same equation; no wonder, as ker L is a vector space and it allows infinitely many basis.

Generally speaking,

Any linear ODE allows infinitely many fundamental systems of solutions.

IS THE CONVERSE TRUE?

The answer to this question is given by the

Theorem 2.3. *To a given fundamental system* $\{y_j\}_{j=\overline{1,n}}$ *it corresponds a unique homogeneous linear differential equation of the form* (2.2.2) *(with the coefficient of* $y^{(n)}$ *equal to* 1*)*.

*Proof. Let $y \in \ker L$. As $\{y_j\}_{j=\overline{1,n}} \subset \ker L$ is a fundamental system of solutions, it is also a basis in ker L, hence we can find n real constants $c_j, j = \overline{1,n}$ such that

$$y(x) = \sum_{j=1}^{n} c_j y_j(x), \quad x \in I. \qquad (2.2.28)$$

But this means that the functions $\{y_1, y_2, \ldots, y_n, y\}$ form a linearly dependent system, which yields $W[y_1, \ldots, y_n, y] \equiv 0$ on I, according to theorem 2.2.

More precisely,

$$\begin{vmatrix} y_1 & y_2 & \cdots & y_n & y \\ y_1' & y_2' & \cdots & y_n' & y' \\ y_1'' & y_2'' & \cdots & y_n'' & y'' \\ \cdots & \cdots & \cdots & \cdots & \cdots \\ y_1^{(n-2)} & y_2^{(n-2)} & \cdots & y_n^{(n-2)} & y^{(n-2)} \\ y_1^{(n-1)} & y_2^{(n-1)} & \cdots & y_n^{(n-1)} & y^{(n-1)} \\ y_1^{(n)} & y_2^{(n)} & \cdots & y_n^{(n)} & y^{(n)} \end{vmatrix} = 0, \quad x \in I. \qquad (2.2.29)$$

By developing the determinant in the left side with respect to the last column, we get a linear ODE in y. It is of nth order, because the coefficient of $y^{(n)}$ is, in fact, the determinant

$$\begin{vmatrix} y_1 & y_2 & \cdots & y_n \\ y_1' & y_2' & \cdots & y_n' \\ \cdots & \cdots & \cdots & \cdots \\ y_1^{(n-1)} & y_2^{(n-1)} & \cdots & y_n^{(n-1)} \end{vmatrix} \equiv W[y_1, y_2, \ldots, y_n], \qquad (2.2.30)$$

which coincides with the Wronskian of the system $\{y_j\}_{j=\overline{1,n}}$ and it does not vanish, because the system $\{y_j\}_{j=\overline{1,n}}$ is fundamental.

By further developing the determinant (2.2.29), we find the coefficient of $y^{(n-1)}$:

$$-\begin{vmatrix} y_1 & y_2 & \cdots & y_n \\ y_1' & y_2' & \cdots & y_n' \\ y_1'' & y_2'' & \cdots & y_n'' \\ \cdots & \cdots & \cdots & \cdots \\ y_1^{(n-2)} & y_2^{(n-2)} & \cdots & y_n^{(n-2)} \\ y_1^{(n)} & y_2^{(n)} & \cdots & y_n^{(n)} \end{vmatrix} \equiv -\frac{\mathrm{d}}{\mathrm{d}x} W\left[y_1, y_2, \ldots, y_n\right]. \qquad (2.2.31)$$

Hence, the equation which allows $\{y_j\}_{j=\overline{1,n}}$ as a fundamental system is of the form

$$W(x) y^{(n)} - \frac{\mathrm{d}}{\mathrm{d}x} W(x) y^{(n-1)} + \ldots = 0. \qquad (2.2.32)$$

Dividing by the coefficient of $y^{(n)}$, which is not null, we obtain the required equation:

$$y^{(n)} - \frac{1}{W(x)} \cdot \frac{\mathrm{d}}{\mathrm{d}x} W(x) y^{(n-1)} + \ldots = 0. \qquad (2.2.33)$$

Comparing this equation with the general form (2.2.2), it follows that

$$p_1(x) = -\frac{1}{W(x)} \cdot \frac{\mathrm{d}}{\mathrm{d}x} W(x), \qquad (2.2.34)$$

or, by integrating once,

$$\ln\left|W\left[y_1, y_2, \ldots, y_n\right]\right| = -\int p_1(x)\,\mathrm{d}x + \ln|C|. \qquad (2.2.35)$$

Passing to exponentials, we obtain **Liouville's formula**, i.e.

$$\boxed{W[y_1, y_2, \ldots, y_n] = C \cdot e^{-\int p_1(x)dx}}.$$ (2.2.36)

UNIQUENESS

As $\dim \ker L = n$, it follows that $n+1$ solutions of the nth order homogeneous linear ODE we are searching for are *linearly dependent*.

Suppose that to the fundamental system $\{y_j\}_{j=\overline{1,n}}$ there correspond two distinct homogeneous linear equations, let them be

$$\begin{aligned} L_1 y &\equiv y^{(n)} + p_1(x)y^{(n-1)} + \ldots + p_{n-1}(x)y' + p_n(x)y = 0, \\ L_2 y &\equiv y^{(n)} + q_1(x)y^{(n-1)} + \ldots + q_{n-1}(x)y' + q_n(x)y = 0. \end{aligned}$$ (2.2.37)

If we subtract the two equations one from another, we obtain

$$(p_1 - q_1)y^{(n-1)} + (p_2 - q_2)y^{(n-2)} + \ldots + (p_{n-1} - q_{n-1})y' + \\ + (p_n - q_n)y = 0.$$ (2.2.38)

As $y_j, j = \overline{1,n}$ satisfy both the ODEs from (2.2.37), it follows that they also satisfy (2.2.38). If $p_1 \neq q_1$, then the order of this ODE is, obviosuly, $(n-1)$. But $\{y_j\}_{j=\overline{1,n}}$ is a fundamental system, hence (2.2.38), a $(n-1)$-order ODE, would allow n linearly independent solutions. **This is a contradiction!**

Therefore $p_1(x) \equiv q_1(x), \forall x \in I$.

In the same way, we can prove that $p_k(x) \equiv q_k(x), \forall x \in I$, for any $k = \overline{2,n}$. ∎

Examples

1. Consider the fundamental system $\{e^x, e^{-x}\}$. Find the correspondent ODE.

a) The order of this equation is 2. We have previously computed $W[e^x, e^{-x}] = -2$, thus proving that $\{e^x, e^{-x}\}$ is a linearly independent system.

b) If y is an arbitrary solution of the equation, then the system $\{e^x, e^{-x}, y\}$ is linearly dependent, therefore its Wronskian vanishes:

$$W[e^x, e^{-x}, y] \equiv 0 \Rightarrow \begin{vmatrix} e^x & e^{-x} & y \\ e^x & -e^{-x} & y' \\ e^x & e^{-x} & y'' \end{vmatrix} = 0. \qquad (2.2.39)$$

Developing this determinant with respect to the last column, it results

$$y'' \begin{vmatrix} e^x & e^{-x} \\ e^x & -e^{-x} \end{vmatrix} - y' \begin{vmatrix} e^x & e^{-x} \\ e^x & e^{-x} \end{vmatrix} + y \begin{vmatrix} e^x & -e^{-x} \\ e^x & e^{-x} \end{vmatrix} = 0. \qquad (2.2.40)$$

Finally, after simplifying with -2, the required equation is

$$\boxed{Ly \equiv y'' - y = 0}.$$

2. Find the ODE allowing the fundamental system $\{\cos x, \sin x\}$.

a) There are two functions in the fundamental system, therefore the order of the required equation is 2.

We verify the linear independence:

$$W[\sin x, \cos x] = \begin{vmatrix} \sin x & \cos x \\ \cos x & -\sin x \end{vmatrix} = -1 \neq 0. \qquad (2.2.41)$$

b) The required equation is given by the Wronskian

$$W[\sin x, \cos x, y] = 0 \Rightarrow \begin{vmatrix} \sin x & \cos x & y \\ \cos x & -\sin x & y' \\ -\sin x & -\cos x & y'' \end{vmatrix} = 0. \qquad (2.2.42)$$

By computing this determinant with respect to the last column, we get

$$y'' \begin{vmatrix} \sin x & \cos x \\ \cos x & -\sin x \end{vmatrix} - y' \begin{vmatrix} \sin x & \cos x \\ -\sin x & -\cos x \end{vmatrix} + \\ + y \begin{vmatrix} \cos x & -\sin x \\ -\sin x & -\cos x \end{vmatrix} = 0. \qquad (2.2.43)$$

After division by $W[\sin x, \cos x] = -1 \neq 0$, the following ODE is obtained:

$$\boxed{Ly \equiv y'' + y = 0}. \qquad (2.2.44)$$

2.3. LINEAR NON-HOMOGENEOUS ODES OF n^{TH} ORDER

Consider again the non-homogeneous linear differential equation

$$\begin{aligned} Ly &\equiv y^{(n)} + p_1(x) y^{(n-1)} + \ldots + p_{n-1}(x) y' + p_n(x) y = \\ &= f(x), \end{aligned} \qquad (2.3.1)$$

whith $p_j, f \in C^0(I)$.

In order to solve it easier, let us present several mathematical issues of great importance.

I. *If Y is a particular solution of the non-homogeneous ODE (2.3.1), and if z is the general solution of the associated homogeneous ODE $Ly = 0$, then* **the general solution** *of the non-homogeneous ODE is*

$$y = Y + z. \qquad (2.3.2)$$

Proof. Let us make the change of function $y = Y + z$, z being the new unknown function. We introduce it in (2.3.1) and, taking into account that L is linear, it results that

$$\begin{cases} Ly = L(Y+z) = LY + Lz \\ = f + Lz \\ Ly = f \end{cases} \Rightarrow f = f + Lz \Rightarrow Lz = 0, \qquad (2.3.3)$$

hence

$$z \in \ker L. \qquad (2.3.4)$$

II. *Suppose that the free term f of the equation (2.3.1) is a sum of the form*

$$f = f_1 + f_2 + \ldots + f_k, \qquad (2.3.5)$$

and let Y_j be some particular solutions corresponding to each f_j, i.e.,

$$LY_j = f_j, \ j = \overline{1,k}. \qquad (2.3.6)$$

Then

$$Y = \sum_{j=1}^{k} Y_j \qquad (2.3.7)$$

is a particular solution of the non-homogeneous equation $LY = f$.

The proof is done by direct computation. Taking into account (2.3.6), we get

$$LY = L\left(\sum_{j=1}^{k} Y_j\right) \underset{L \text{ linear}}{=} \sum_{j=1}^{k} LY_j = \sum_{j=1}^{k} f_j = f. \qquad (2.3.8)$$

III. *If we know a fundamental system of solutions of the ODE* (2.2.26), *then, by using* **the variation of parameters**, *we can find a particular solution of the non-homogeneous equation* (2.3.1).

Proof. For a better understanding, let us take $n = 3$, which is easily generalized to an arbitrary n.

Consider therefore the equation

$$\begin{aligned}Ly &\equiv y''' + p_1(x)y'' + p_2(x)y' + p_3(x)y = \\ &= f(x).\end{aligned} \qquad (2.3.9)$$

and let $\{y_1, y_2, y_3\} \subset \ker L$ be a fundamental system of solutions.

Then $Ly_j = 0, j = \overline{1,3}$. According to statement **I**, the general solution of (2.3.9) is the function $y = Y + z$, where z is the general solution of the associated to (2.3.9) homogeneous ODE:

$$Ly \equiv y''' + p_1(x)y'' + p_2(x)y' + p_3(x)y = 0. \qquad (2.3.10)$$

As $\{y_1, y_2, y_3\}$ is a fundamental system, it is also a basis in $\ker L$, hence z is expressed as a linear combination of the functions of the system:

$$z(x) = c_1 y_1(x) + c_2 y_2(x) + c_3 y_3(x). \tag{2.3.11}$$

As in the case of first-order linear ODEs, we search for Y of the form

$$Y(x) = c_1(x) y_1 + c_2(x) y_2 + c_3(x) y_3. \tag{2.3.12}$$

Introducing this into the equation, we get

$$\begin{array}{l|l}
Y = c_1 y_1 + c_2 y_2 + c_3 y_3 & p_3(x) \\
Y' = c_1 y_1' + c_2 y_2' + c_3 y_3' + \underbrace{c_1' y_1 + c_2' y_2 + c_3' y_3}_{=0} & p_2(x) \\
Y'' = c_1 y_1'' + c_2 y_2'' + c_3 y_3'' + \underbrace{c_1' y_1' + c_2' y_2' + c_3' y_3'}_{=0} & p_1(x) \\
Y''' = c_1 y_1''' + c_2 y_2''' + c_3 y_3''' + c_1' y_1'' + c_2' y_2'' + c_3' y_3'' & 1
\end{array} \quad + \tag{2.3.13}$$

$$L Y = \underbrace{c_1 L y_1}_{=0} + \underbrace{c_2 L y_2}_{=0} + \underbrace{c_3 L y_3}_{=0} + c_1' y_1'' + c_2' y_2'' + c_3' y_3'' = f.$$

Hence c_1', c_2', c_3' satisfy the linear algebraic system

$$\begin{aligned}
c_1' y_1 + c_2' y_2 + c_3' y_3 &= 0, \\
c_1' y_1' + c_2' y_2' + c_3' y_3' &= 0, \\
c_1' y_1'' + c_2' y_2'' + c_3' y_3'' &= f.
\end{aligned} \tag{2.3.14}$$

The associated determinant is, in fact, the Wronskian of the fundamental system, therefore it is non-zero on I:

$$W[y_1, y_2, y_3] \equiv \begin{vmatrix} y_1(x) & y_2(x) & y_3(x) \\ y_1'(x) & y_2'(x) & y_3'(x) \\ y_1''(x) & y_2''(x) & y_3''(x) \end{vmatrix} \neq 0, \quad x \in I. \quad (2.3.15)$$

Consequently, the system (2.3.14) allows a unique solution. Let

$$c_j' = \varphi_j(x), j = \overline{1,3}, \quad (2.3.16)$$

be this solution. Integrating it, we obtain

$$c_j = \int \varphi_j(x) dx, j = \overline{1,3}. \quad (2.3.17)$$

IV. Conclusion: *If one knows a fundamental system of solutions of the non-homogeneous ODE (2.3.1), then its general solution is obtained by* **quadratures.**

Proof. If $\{y_j\}_{j=\overline{1,n}}$ is a fundamental system of solutions, it results that the general solution of the associated to (2.3.1) homogeneous ODE, i.e., $Ly = 0$, is

$$y_{\text{homogeneous}} = \sum_{j=1}^{n} c_j y_j, \quad x \in I. \quad (2.3.18)$$

Then

1. According to statement **III**, a particular solution Y of (2.3.1) is

$$Y = \sum_{j=1}^{n} \left(\int \varphi_j(x) dx \right) \cdot y_j, \quad (2.3.19)$$

where $c_j' = \varphi_j$ are solutions of the algebraic system

$$\begin{aligned}c'_1 y_1 + c'_2 y_2 + \ldots + c'_n y_n &= 0,\\ c'_1 y'_1 + c'_2 y'_2 + \ldots + c'_n y'_n &= 0,\\ &\cdots\cdots\cdots\cdots\cdots\cdots\cdots\cdots\cdots\cdots\cdots\\ c'_1 y''_1 + c'_2 y''_2 + \ldots + c'_n y''_n &= f.\end{aligned} \quad (2.3.20)$$

2. By statement **I**, $y = Y + z$, hence the general solution of the ODE (2.3.1) is

$$y(x) = \sum_{j=1}^{n} c_j y_j + \sum_{j=1}^{n} \left(\int \varphi_j(x) \mathrm{d}x \right) \cdot y_j, \quad (2.3.21)$$

where $\varphi_j = c'_j$, $j = \overline{1,n}$ satisfy the system (2.3.20).

Example. Solve the equation

$$Ly \equiv y'' - y = \mathrm{e}^{2x}. \quad (2.3.22)$$

STEP 1. The associated homogeneous equation is

$$Ly \equiv y'' - y = 0, \quad (2.3.23)$$

and it allows the fundamental system of solutions $\{\mathrm{e}^x, \mathrm{e}^{-x}\}$ (they have been found in the previous examples).

Hence, the general solution of the homogeneous ODE is

$$y_{\text{homogeneous}} = c_1 \mathrm{e}^x + c_2 \mathrm{e}^{-x}. \quad (2.3.24)$$

STEP 2. We search for a particular solution of the non-homogeneous ODE of the form

$$Y = c_1(x)\mathrm{e}^x + c_2(x)\mathrm{e}^{-x}. \quad (2.3.25)$$

By introducing it into the equation, we find

$$Y = c_1(x)e^x + c_2(x)e^{-x}$$
$$Y' = c_1(x)e^x - c_2(x)e^{-x} + \underbrace{c_1'(x)e^x + c_2'(x)e^{-x}}_{=0}$$
$$Y'' = c_1(x)e^x + c_2(x)e^{-x} + c_1'(x)e^x - c_2'(x)e^{-x}$$
$$\begin{vmatrix} -1 \\ 0 \\ 1 \end{vmatrix} + \qquad (2.3.26)$$
$$Ly = \quad / \quad + \quad / \quad + c_1'(x)e^x - c_2'(x)e^{-x} = e^{2x}.$$

Therefore, we must solve the system

$$\begin{aligned} c_1'e^x + c_2'e^{-x} &= 0, \\ c_1'e^x - c_2'e^{-x} &= e^{2x}. \end{aligned} \qquad (2.3.27)$$

The associated determinant coincides with the Wronskian:

$$\Delta = W\left[e^x, e^{-x}\right] = -2. \qquad (2.3.28)$$

We obtain the unique solution

$$c_1' = -\frac{1}{2}\begin{vmatrix} 0 & e^{-x} \\ e^{2x} & -e^{-x} \end{vmatrix} = \frac{1}{2}e^x,$$
$$c_2' = -\frac{1}{2}\begin{vmatrix} e^x & 0 \\ e^x & e^{2x} \end{vmatrix} = -\frac{1}{2}e^{3x}. \qquad (2.3.29)$$

By integration, we obtain

$$c_1 = \frac{1}{2}e^x, \, c_2 = -\frac{1}{6}e^{3x}; \qquad (2.3.30)$$

thus, according to (2.3.25), the particular solution Y is expressed as

$$Y = \frac{1}{2}e^x \cdot e^x - \frac{1}{6}e^{3x} \cdot e^{-x}, \qquad (2.3.31)$$

whence it follows that

$$Y = \frac{1}{3}e^{2x}. \tag{2.3.32}$$

Consequently, the general solution of the non-homogeneous ODE (2.3.22) is

$$\boxed{y(x) = c_1 e^x + c_2 e^{-x} + \frac{1}{3}e^{2x}}. \tag{2.3.33}$$

Remark. In the case of constant coefficients p_j and free terms expressed by elementary functions, the particular solution Y is easier found by shaping it into the free term form.

Example. For the equation (2.3.22), we can search for an Y of the form $Y = k e^{2x}$. By introducing this expression into the equation, we obtain

$$\begin{array}{c} Y = k e^{2x} \quad \begin{vmatrix} -1 \\ 0 \\ 1 \end{vmatrix} + \\ Y' = 2k e^{2x} \\ Y'' = 4k e^{2x} \\ \hline LY = (4k - k)e^{2x} = 3k e^{2x}. \end{array} \tag{2.3.34}$$

Therefore, we must have

$$3k e^{2x} = e^{2x} \implies k = \frac{1}{3}; \tag{2.3.35}$$

we obtain $\boxed{Y = \frac{1}{3}e^{2x}}$, a particular solution easier obtained than by using the variation of parameters.

2.4. LINEAR ODEs OF ORDER N, WITH CONSTANT COEFFICIENTS

The general form of these equations is

$$Ly \equiv a_0 y^{(n)} + a_1 y^{(n-1)} + a_2 y^{(n-2)} + \ldots + a_{n-1} y' + a_n y = \\ = f(x), \qquad (2.4.1)$$

where $a_k \in \mathfrak{R}$, $k = \overline{0,n}$, $a_0 \neq 0$.

In the previous section we saw that, once we know a fundamental system of solutions, the general solution of a non-homogeneous linear ODE is completely obtained by quadratures. In the case of ODEs with constant coefficients, we can always determine a fundamental system of solutions.

2.4.1. LINEAR HOMOGENEOUS ODEs

Let

$$Ly \equiv a_0 y^{(n)} + a_1 y^{(n-1)} + a_2 y^{(n-2)} + \ldots + a_{n-1} y' + a_n y = 0, \\ a_j \in \mathfrak{R}, j = \overline{0,n}, \qquad (2.4.2)$$

be the homogeneous equation associated to (2.4.1). The operator L, which is defined by the left side of this equation, is linear, as proved in section 2.1.

We remind now several facts already proved and discussed in the previous sections.

The kernel of the operator L is

$$\ker L = \{ y \in C^n(\mathfrak{R}) \mid Ly = 0 \}, \qquad (2.4.3)$$

therefore

***The set of the solutions of the ODE** (2.4.2) **coincides with** ker L.*

As we have shown in section 2.2, the dimension of ker L is n. Consequently,

In order to solve a linear n^{th} order ODE, we must find a basis in ker L.

We remind that a basis of a n-dimensional vector space is a set containing n linearly independent elements of the space. Let $\{y_1, y_2, \ldots, y_n\}$ be a basis in ker L. Then the general solution of the ODE (2.4.2) is written as a linear combination with arbitrary coefficients of the elements of the basis, i.e.,

$$y(x) = c_1 y_1(x) + c_2 y_2(x) + \ldots + c_n y_n(x). \qquad (2.4.4)$$

METHOD OF SOLVING

In the case of constant coefficients, we search for solutions of the exponential form $y = e^{rx}$, using **Leonhard Euler's idea**. We differentiate it step by step and we introduce it into the equation:

$$\begin{array}{rl} a_n \times & y = e^{rx} \\ a_{n-1} \times & y' = r e^{rx} \\ a_{n-2} \times & y'' = r^2 e^{rx} \\ \ldots \\ a_1 \times & y^{(n-1)} = r^{n-1} e^{rx} \\ a_0 \times & y^{(n)} = r^n e^{rx} \\ \hline Ly = e^{rx}\left(a_0 r^n + a_1 r^{n-1} + \ldots a_{n-1} r + a_n\right) = 0, \end{array} \qquad (2.4.5)$$

thus, in order for e^{rx} to be a solution, we must have

$$\boxed{a_0 r^n + a_1 r^{n-1} + \ldots a_{n-1} r + a_n = 0}. \tag{2.4.6}$$

The equation (2.4.6) is called *the characteristic equation*. It always allows n roots in the complex field, which is algebraically closed. Let $r_1, r_2, \ldots, r_n$ be these roots. We have several posibilities:

A. Real and distinct roots. In this case, a basis of ker L is formed by the functions $e^{r_1 x}, e^{r_2 x}, \ldots, e^{r_n x}$, according to the correspondence

$$\begin{array}{cccc} r_1 & r_2 & r_3 & r_n \\ \downarrow & \downarrow & \downarrow & \cdots & \downarrow \\ e^{r_1 x} & e^{r_2 x} & e^{r_3 x} & & e^{r_n x}, \end{array} \tag{2.4.7}$$

hence the general solution of the ODE (2.4.2) is

$$y(x) = c_1 e^{r_1 x} + c_2 e^{r_2 x} + \ldots + c_n e^{r_n x}. \tag{2.4.8}$$

B. Complex conjugate roots. Let $r_1 = a + ib$. Then the characteristic equation, which has real coefficients, also allows $r_2 = a - ib$ as a root. For simplification, let us assume that the other roots are real and distinct. To preserve the real framework, we shall replace $e^{(a+ib)x}, e^{(a-ib)x}$ with their real linear combinations, by using Euler's formulas (see e.g. [2][3][5][10]):

$$\begin{aligned} e^{ax} \cos bx &= e^{ax} \frac{e^{ibx} + e^{-ibx}}{2}, \\ e^{ax} \sin bx &= e^{ax} \frac{e^{ibx} - e^{-ibx}}{2i}. \end{aligned} \tag{2.4.9}$$

Then the diagram (2.4.7) becomes

$$\begin{matrix} r_1 & r_2 & r_3 & r_n \\ \downarrow & \downarrow & \downarrow & \downarrow \\ e^{ax}\cos bx & e^{ax}\sin bx & e^{r_3 x} & \cdots & e^{r_n x}, \end{matrix} \qquad (2.4.10)$$

and the general solution of the ODE (2.4.2) is

$$y(x) = e^{ax}(c_1 \cos bx + c_2 \sin bx) + c_3 e^{r_3 x} \ldots + c_n e^{r_n x}. \qquad (2.4.11)$$

C. Multiple roots. Unlike the previous cases, this one needs more explanations. We cannot directly use the diagram (2.4.7), because we would obviously obtain a linearly dependent system.

So, let us firstly consider the second-order equation

$$Ly \equiv ay'' + by' + cy = 0. \qquad (2.4.12)$$

Suppose that its characteristic equation

$$ar^2 + br + c = 0, \qquad (2.4.13)$$

allows the real roots r_1, r_2, which are significantly close, but distinct.

Then we can use the diagram (2.4.7), which, in this case, becomes

$$\begin{matrix} r_1 & r_2 \\ \downarrow & \downarrow \\ e^{r_1 x} & e^{r_2 x}. \end{matrix} \qquad (2.4.14)$$

If $r_2 \to r_1$, then the diagram does not give a fundamental system. In order to eliminate this shortcoming, we can replace $e^{r_2 x}$ with the linear combination

$$\frac{e^{r_2 x} - e^{r_1 x}}{r_2 - r_1}, \tag{2.4.15}$$

which is, obviously, also a solution of the ODE (2.4.12). Passing to the limit for $r_2 \to r_1$, we obtain

$$\lim_{r_2 \to r_1} \frac{e^{r_2 x} - e^{r_1 x}}{r_2 - r_1} = \lim_{r_2 \to r_1} \frac{\frac{d}{dr_2}(e^{r_2 x} - e^{r_1 x})}{\frac{d}{dr_2}(r_2 - r_1)} = \tag{2.4.16}$$

$$= \lim_{r_2 \to r_1} \frac{x e^{r_2 x}}{1} = x e^{r_1 x}.$$

This means that, if $r_2 = r_1$, we can consider, for the equation (2.4.12), the diagram

$$\begin{array}{cc} r_1 & r_1 \\ \downarrow & \downarrow \\ e^{r_1 x} & x e^{r_1 x}. \end{array} \tag{2.4.17}$$

Indeed, the two functions specified in the diagram are solutions of the equation and they are linearly independent, because their Wronskian

$$W\left[e^{r_1 x}, x e^{r_1 x}\right] = \begin{vmatrix} e^{r_1 x} & x e^{r_1 x} \\ r_1 e^{r_1 x} & x r_1 e^{r_1 x} + e^{r_1 x} \end{vmatrix} =$$

$$= e^{2 r_1 x} \begin{vmatrix} 1 & x \\ r_1 & x r_1 + 1 \end{vmatrix} = e^{2 r_1 x} \tag{2.4.18}$$

is non-zero. The general solution of the ODE (2.4.12) is

$$y(x) = c_1 e^{r_1 x} + c_2 x e^{r_1 x}, \tag{2.4.19}$$

or else

$$y(x) = e^{r_1 x}(c_1 + c_2 x). \qquad (2.4.20)$$

Consider now the general case. For simplicity, we assume that r_1 is a multiple root of mth order of the characteristic equation (2.4.6), and the other roots $r_{m+1}, r_{m+2}, \ldots, r_n$ are all of them real and distinct.

According to the previous remark for the case $n = 2$, we infer that, in this case, the diagram (2.4.7) becomes

$$\begin{array}{ccccc} r_1 & r_1 & r_1 & r_{m+1} & r_n \\ \downarrow & \downarrow & \cdots \downarrow & \downarrow & \cdots \downarrow \\ e^{r_1 x} & x e^{r_1 x} & x^{m-1} e^{r_1 x} & e^{r_{m+1} x} & e^{r_n x}. \end{array} \qquad (2.4.21)$$

and, therefore, the general solution of the ODE (2.4.2) is

$$y(x) = e^{r_1 x}\left(c_1 + c_2 x + \ldots c_m x^{m-1}\right) + c_3 e^{r_3 x} \ldots + c_n e^{r_n x}. \qquad (2.4.22)$$

CONCLUSION: For ODEs with constant coefficients, we can always set up a fundamental system of solutions, expressed in terms of elementary functions.

Examples. Find the general solution of the following ordinary differential equations:

a) $Ly \equiv y'' - 3y' + 2y = 0$.

This is a second-order linear ODE, with constant coefficients.

The dimension of ker L is 2. Searching for solutions of exponential form, i.e., $y = e^{rx}$, we deduce that r must satisfy the characteristic equation

$$r^2 - 3r + 2 = 0,$$

which allows the real and distinct roots $r_1 = 1, r_2 = 2$. This corresponds to case **A**.

According to formula (2.4.8), the general solution of **a)** is

$$\boxed{y = c_1 e^x + c_2 e^{2x}}.$$

b) $Ly \equiv y'' + y = 0$.

This is also a second-order linear ODE with constant coefficients.

The dimension of ker L is 2. Serching for solutions of exponential form $y = e^{rx}$, we infer that r must satisfy the characteristic equation

$$r^2 + 1 = 0,$$

which allows the pure imaginary roots $r_1 = +\mathrm{i}, r_2 = -\mathrm{i}$.

According to formula (2.4.11), the equation **b)** allows the general solution

$$\boxed{y = c_1 \cos x + c_2 \sin x}.$$

c) $Ly \equiv y'' + 2y' + y = 0$.

This is a second-order linear ODE with constant coefficients.

The dimension of ker L is 2. Looking for solutions of exponential form $y = e^{rx}$, it results that r must satisfy the characteristic equation

$$r^2 + 2r + 1 = 0,$$

which allows the double root $r = -1$. According to formula (2.4.22), the general solution of *c)* is

$$\boxed{y = (c_1 + c_2 x)e^{-x}}.$$

APPLICATION: THE HARMONIC OSCILLATOR

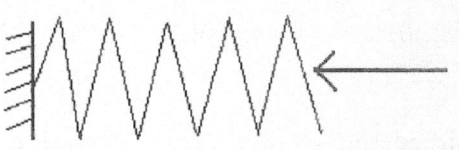

Figure 2.1. The harmonic oscillator

The figure 2.1 is self explanatory. We firstly set up **the mathematical model associated to this physical phenomenon**.

As previously shown, setting up a model means:
- ♣ to establish **the physical quantity** (or **quantities**), which determines the complete knowledge of the physical phenomenon; this quantity will play the part of the unknown function and
- ♣ to establish **the physical law** (or **laws**), which governs the phenomenon and to give their mathematical expressions.

THE MATHEMATICAL MODEL OF THE HARMONIC OSCILLATOR

1. In this case, the unknown function is **the displacement** $y = y(t)$, with a unique component.

2. The physical law is ***Newton's Law***: The vector sum of the forces **F** on an object is equal to the mass *m* of that object multipled by vector **a** representing the acceleration of the object, i.e.,

$$m\mathbf{a} = \mathbf{F}, \tag{2.4.23}$$

where **a**, **F** have each of them a unique component.

Here, **F** is the elastic force and it is expressed as

$$\mathbf{F} = -ky, \quad k > 0, \tag{2.4.24}$$

It is already known that the acceleration a is the second derivative of the displacement with respect to time:

$$\mathbf{a} = \frac{d^2 y}{dt^2}. \tag{2.4.25}$$

We shall consider the following notation for the derivatives with respect to time, usual in mechanics:

$$\frac{d^2 y}{dt^2} \equiv \ddot{y}. \tag{2.4.26}$$

Therefore

$$m\ddot{y} = -ky. \tag{2.4.27}$$

Obviously, as $\frac{k}{m} > 0$, we can put $\frac{k}{m} = \omega^2$. Finally, the mathematical model of the harmonic oscillator is given by

$$\boxed{Ly \equiv \ddot{y} + \omega^2 y = 0}. \tag{2.4.28}$$

This is a homogeneous linear ODE with constant coefficients. In order to find a fundamental system of solutions, we search for y of exponential form, i.e., $y = e^{\alpha t}$. Differentiating and introducing it into the equation yields:

$$\begin{array}{c|c} \omega^2 & y = e^{\alpha t} \\ 0 & \dot{y} = \alpha e^{\alpha t} \\ 1 & \ddot{y} = \alpha^2 e^{\alpha t} \end{array} + \Rightarrow Ly = (\alpha^2 + \omega^2)e^{\alpha t} = 0. \qquad (2.4.29)$$

We thus obtained the characteristic equation

$$\alpha^2 + \omega^2 = 0, \qquad (2.4.30)$$

with the roots $\alpha_{1,2} = \pm i\omega$. The equation has the following fundamental system:

$$\begin{array}{cc} i\omega & -i\omega \\ \downarrow & \downarrow \\ e^{i\omega t} & e^{-i\omega t}. \end{array} \qquad (2.4.31)$$

In order to keep the real framework, let us use **Euler's formulas** (see point **B** of this section). We have:

$$\begin{aligned} e^{i\alpha} &= \cos\alpha + i\sin\alpha, \\ e^{-i\alpha} &= \cos\alpha - i\sin\alpha, \end{aligned} \qquad (2.4.32)$$

hence

$$\frac{e^{i\alpha} + e^{-i\alpha}}{2} = \cos\alpha, \quad \frac{e^{i\alpha} - e^{-i\alpha}}{2i} = \sin\alpha. \qquad (2.4.33)$$

Instead of exponential functions with complex coefficients, we can therefore use the combinations

$$\frac{e^{i\omega t} + e^{-i\omega t}}{2} = \cos \omega t,$$
$$\frac{e^{i\omega t} - e^{-i\omega t}}{2} = \sin \omega t, \qquad (2.4.34)$$

as they are also solutions and, moreover, they form a fundamental system.

Indeed

$$W[\cos \omega t, \sin \omega t] = \begin{vmatrix} \cos \omega t & \sin \omega t \\ -\omega \sin \omega t & \omega \cos \omega t \end{vmatrix} = \omega \neq 0. \qquad (2.4.35)$$

The general solution of the ODE (2.4.28) is

$$y(t) = c_1 \cos \omega t + c_2 \sin \omega t. \qquad (2.4.36)$$

Instead of the arbitrary constants c_1, c_2, we shall consider two other constants A and δ, which are also arbitrary.

Putting $A = \sqrt{c_1^2 + c_2^2}$, it follows that

$$y(t) = A\left(\frac{c_1}{\sqrt{c_1^2 + c_2^2}} \cos \omega t + \frac{c_2}{\sqrt{c_1^2 + c_2^2}} \sin \omega t\right). \qquad (2.4.37)$$

The constants in the brackets are, obviously, subunitary and the sum of their squares is 1, so we can put

$$\frac{c_1}{\sqrt{c_1^2 + c_2^2}} = \cos \delta, \quad \frac{c_2}{\sqrt{c_1^2 + c_2^2}} = \sin \delta, \qquad (2.4.38)$$

where

$$\delta = \arctan\frac{c_1}{c_2}. \qquad (2.4.39)$$

Finally, the general solution of the harmonic oscillator equation is expressed as follows

$$\boxed{y(t) = A\cos(\omega t - \delta)}. \qquad (2.4.40)$$

PHYSICAL INTERPRETATION

In the figure 2.2 we draw the graph of the function (2.4.40), where

- A is ***the amplitude of motion***,
- ω is ***the frequency of motion***,
- δ is ***the phase of motion***.

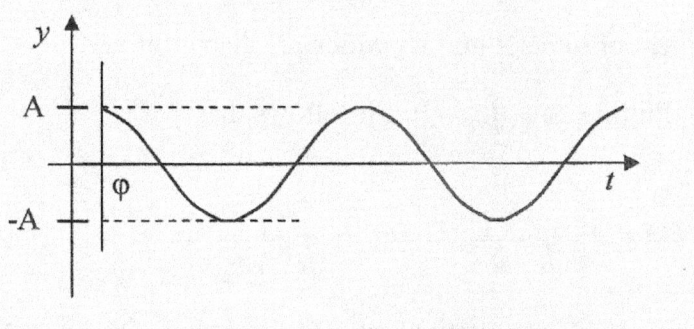

Figure 2.2. The geometrical representation of the motion of the harmonic oscillator

By setting to zero the argument of the cosine, we obtain the moment $\varphi = \dfrac{\delta}{\omega}$, which corresponds to the amplitude A.

2.4.2. DIFFERENTIAL POLYNOMIAL

Consider again the equation (2.4.1).

We notice that it can be also written as:

$$Ly \equiv a_0 \frac{d^n}{dx^n} y + a_1 \frac{d^{n-1}}{dx^{n-1}} y + \ldots + a_{n-1} \frac{d}{dx} y + a_n y \quad (2.4.41)$$
$$= f(x).$$

Let us denote the differential operator by D:

$$D \equiv \frac{d}{dx}. \quad (2.4.42)$$

Then

$$\frac{d^k}{dx^k} = D^k, \quad (2.4.43)$$

and thus the operator L is written in terms of D:

$$Ly \equiv a_0 D^n y + a_1 D^{n-1} y + \ldots + a_{n-1} Dy + a_n Ey$$
$$= f(x). \quad (2.4.44)$$

In (2.4.44), we used the notation E for the identity operator, i.e.,

$$Ey = y. \quad (2.4.45)$$

We can put (2.4.44) in the form

$$Ly \equiv \left(a_0 D^n + a_1 D^{n-1} + \ldots + a_{n-1} D + a_n E \right) y$$
$$= f(x). \quad (2.4.46)$$

The above operator in the brackets is, formally, a polynomial of degree n in D.

This is what we call *a **differential polynomial***.

We shall use the following notation for the differential polynomial:

$$P_n(D) \equiv a_0 D^n + a_1 D^{n-1} + \ldots + a_{n-1} D + a_n E. \qquad (2.4.47)$$

It follows that the ODE (2.4.1) can be shaped as:

$$Ly \equiv P_n(D) y = f(x). \qquad (2.4.48)$$

Remark. By replacing D with r in (2.4.47) and the higher derivatives with degrees, we obtain the characteristic polynomial associated to the differential equation.

USEFUL COMPUTATION FORMULAS

I. Let us apply the differential polynomial to the exponential function

$$y = e^{\alpha x}. \qquad (2.4.49)$$

Taking into account that

$$D(e^{\alpha x}) = \alpha e^{\alpha x}, \quad D^k(e^{\alpha x}) = \alpha^k e^{\alpha x}, \qquad (2.4.50)$$

we obtain

$$\begin{aligned} P_n(D)(e^{\alpha x}) &= \\ &= (a_0 D^n + a_1 D^{n-1} + \ldots + a_{n-1} D + a_n E) e^{\alpha x} = \\ &= a_0 D^n e^{\alpha x} + a_1 D^{n-1} e^{\alpha x} + \ldots + a_{n-1} D e^{\alpha x} + a_n E e^{\alpha x} = . \qquad (2.4.51)\\ &= a_0 \alpha^n e^{\alpha x} + a_1 \alpha^{n-1} e^{\alpha x} + \ldots + a_{n-1} \alpha e^{\alpha x} + a_n e^{\alpha x} = \\ &= (a_0 \alpha^n + a_1 \alpha^{n-1} + \ldots + a_{n-1} \alpha + a_n) e^{\alpha x}, \end{aligned}$$

therefore

$$\boxed{P_n(D)(e^{\alpha x}) = P_n(\alpha)e^{\alpha x}}. \qquad (2.4.52)$$

This important formula is very useful in practice. As a matter of fact, we used it also in the previous paragraphs, in connection with the characteristic equation.

II. We can prove another useful formula of computation, true for any differential polynomial.

Lemma 2.1. *If* $u, v \in C^n(I)$, *then*

$$\boxed{\begin{aligned} P_n(D)(uv) &= \\ = u P_n(D)v &+ \frac{1}{1!} u' P_n'(D)v + \frac{1}{2!} u'' P_n''(D)v + \\ + \ldots &+ \frac{1}{(n-1)!} u^{(n-1)} P_n^{(n-1)}(D)v + \frac{1}{n!} u^{(n)} P_n^{(n)}(D)v. \end{aligned}} \qquad (2.4.53)$$

* **The proof** is done by using Leibniz's formula

$$\begin{aligned} D^k(uv) &= u D^k v + u' D^{k-1} v + C_n^2 u'' D^{k-2} v + \\ &+ \ldots + C_n^{n-1} u^{(n-1)} D v + v D^n u. \end{aligned} \qquad (2.4.54)$$

Let us prove this for $n = 3$; for arbitrary n, we can easily prove it by induction. Consider therefore the operator

$$P(D) \equiv a D^3 + b D^2 + c D + d E. \qquad (2.4.55)$$

We have

$$\begin{array}{l|l}
D^3(uv) = u D^3 v + C_3^1 D u D^2 v + C_3^2 D^2 u D v + v D^3 u & \times a \\
D^2(uv) = u D^2 v + C_2^1 D u D v + v D^2 u & \times b \\
D(uv) = u D v + v D u & \times c \\
E(uv) = u v & \times d
\end{array} \qquad (2.4.56)$$

$$P(D)(uv) = u\left(aD^3v + bD^2v + cDv + Ev\right) +$$
$$+ Du\left(aC_3^1D^2v + bC_2^1Dv + cEv\right) +$$
$$+ D^2u\left(aC_3^2Dv + bEv\right) + avD^3u.$$

We notice that

$$u\left(aD^3v + bD^2v + cDv + Ev\right) = uP(D)v,$$
$$Du\left(aC_3^1D^2v + bC_2^1Dv + cEv\right) =$$
$$= Du\left(3aD^2v + 2Dv + cv\right) = DuP'(D)v,$$
$$D^2u\left(aC_3^2Dv + bEv\right) = D^2u\left(3aDv + bv\right) = \quad (2.4.57)$$
$$= \frac{1}{2}D^2u\left(6aDv + 2bv\right) = \frac{1}{2}D^2uP''(D)v,$$
$$D^3u\left(av\right) = D^2u\frac{1}{3!}P'''(D)v.$$

Therefore, the formula (2.4.53) is proven for $n = 3$. ∎

2.4.3. NON-HOMOGENEOUS LINEAR ODEs

We saw that in the case of ODEs with constant coefficients one can always determine a fundamental system of closed-form solutions. We only need to find a particular solution of the non-homogeneous equation:

$$Ly \equiv a_0 y^{(n)} + a_1 y^{(n-1)} + a_2 y^{(n-2)} + \ldots + a_{n-1} y' + \quad (2.4.58)$$
$$+ a_n y = f(x).$$

Obviously, we can apply the variation of parameters, but, in the case of constant coefficients, we can use more efficient methods, if the free term is also expressed by elementary functions.

We distinguish several cases:

A. *The free term is a polynomial of order m in x*:

$$f(x) = P_m(x). \qquad (2.4.59)$$

Then

- If $a_n \neq 0$, we search for the particular solution $Y(x)$ of the ODE $LY = P_m(x)$ in the form of a polynomial of the same degree, i.e.,

$$Y(x) = Q_m(x). \qquad (2.4.60)$$

Elementary computation immediately leads to the coefficients of $Q_m(x)$.

Example. Find a particular solution of the ODE

$$Ly \equiv y''' + y'' - y' - y = x^2 + 1. \qquad (2.4.61)$$

Solution. The equation (2.4.61) is a non-homogeneous linear equation with constant coefficients. The free term is a second degree polynomial and $a_3 \equiv -1 \neq 0$. We can search for the particular solution in the form of a second degree polynomial, i.e.,

$$Y(x) \equiv Q_2(x) = ax^2 + bx + c. \qquad (2.4.62)$$

Differentiating and introducing it into the equation, we obtain

$$2a - (2ax + b) - (ax^2 + bx + c) = x^2 + 1, \qquad (2.4.63)$$

or

$$-ax^2 - (2a+b)x + 2a - b - c = x^2 + 1. \qquad (2.4.64)$$

We identify the coefficients of the same powers of x, thus getting

$$a = -1, \quad b = 2, \quad c = -5. \qquad (2.4.65)$$

Hence, the required particular solution is

$$\boxed{Y(x) = -x^2 + 2x - 5}. \qquad (2.4.66)$$

- If $a_n, a_{n-1}, \ldots, a_{n-r}$ $(r < n)$ are all of them null, we search for Y of the form

$$Y(x) = x^{r-1} Q_m(x). \qquad (2.4.67)$$

Example. Find a particular solution of the non-homogeneous linear ODE

$$Ly \equiv y''' + y'' = x + 1. \qquad (2.4.68)$$

Solution. The equation (2.4.68) is an equation with constant coefficients. The free term is a polynomial of degree 1 and $a_3 = 0, a_2 = 0$. Therefore, we search for a particular solution Y of the form

$$Y(x) = x^2 Q_1(x) = x^2 (ax + b). \qquad (2.4.69)$$

By differentiating and introducing it into the equation, we obtain

$$6a + (6ax + 2b) = x + 1, \qquad (2.4.70)$$

or

$$6ax + 6a + 2b = x + 1. \qquad (2.4.71)$$

Identifying the coefficients of the same powers of x, we find

$$a = \frac{1}{6}, \quad b = 0, \qquad (2.4.72)$$

therefore, the required particular solution is

$$\boxed{Y(x) = \frac{1}{6}x^3}. \qquad (2.4.73)$$

B. *The free term is an exponential function*:

$$f(x) = A e^{\alpha x}. \qquad (2.4.74)$$

Here, we distinguish two cases:

- **α *is not a root of the characteristic equation***, therefore $P_n(\alpha) \neq 0$. In this case, we search for a particular solution Y of the form

$$Y(x) = a e^{\alpha x}. \qquad (2.4.75)$$

Differentiating and introducing it in the equation, we obtain

$$a P_n(\alpha) e^{\alpha x} = A e^{\alpha x}, \qquad (2.4.76)$$

whence, by elementary calculation, we deduce

$$a = \frac{A}{P_n(\alpha)}. \qquad (2.4.77)$$

Example. Find a particular solution of the ODE

$$Ly \equiv y'' - 3y' + 2y = e^{3x}. \qquad (2.4.78)$$

Solution. The equation (2.4.78) is a non-homogeneous linear equation with constant coefficients. The free term is of the exponential form (2.4.75), with $\alpha = 3$. The equation is also shaped by using the differential polynomial

$$Ly \equiv \underbrace{\left(D^2 - 3D + 2E\right)}_{P(D)} y = e^{3x}. \qquad (2.4.79)$$

The associated characteristic equation is

$$r^2 - 3r + 2 = 0, \qquad (2.4.80)$$

with the roots

$$r_1 = 1, \quad r_2 = 2; \qquad (2.4.81)$$

none of them coincides with α. Therefore, we search for a particular solution Y of the form

$$Y(x) = a e^{3x}. \qquad (2.4.82)$$

Introducing it in the equation after differentiation, we are lead to

$$P(D)\left(a e^{3x}\right) = a \cdot P\left(e^{3x}\right) = a e^{3x}(9 - 3 \cdot 3 + 2) \\ = 2 a e^{3x}. \qquad (2.4.83)$$

whence we obtain

$$2a e^{3x} = e^{3x} \quad \rightarrow \quad a = \frac{1}{2}. \qquad (2.4.84)$$

Therefore, a particular solution for the ODE (2.4.78) is

$$\boxed{Y(x) = \frac{1}{2}e^{3x}}. \qquad (2.4.85)$$

- **α *is a multiple root of order* m, $m \leq n$**, of the characteristic equation, therefore

$$P_n(\alpha) = 0,\ P_n'(\alpha) = 0,\ P_n''(\alpha) = 0,\ \ldots P_n^{(m-1)}(\alpha) \neq 0, \qquad (2.4.86)$$

but

$$P_n^{(m)}(\alpha) \neq 0. \qquad (2.4.87)$$

In this case, we search for a particular solution Y of the form

$$Y(x) = a x^m e^{\alpha x}. \qquad (2.4.88)$$

Let us differentiate it by using formula (2.4.53), in which we put $u = a x^m$, $v = e^{\alpha x}$. Introducing it into the equation, we obtain

$$a_0 a x^m \underbrace{P_n(\alpha) e^{\alpha x}}_{=0} + a_1 m a x^{m-1} \underbrace{P_n'(\alpha) e^{\alpha x}}_{=0} + \ldots +$$

$$+ \frac{1}{(m-1)!}(m-1)! a x \cdot a_{m-1} \underbrace{P_n^{(m-1)}(\alpha) e^{\alpha x}}_{=0} + \qquad (2.4.89)$$

$$+ \frac{1}{m!} m! a \cdot a_m P_n^{(m)}(\alpha) e^{\alpha x} = A e^{\alpha x},$$

whence, by elementary calculation, we get

$$a = \frac{A}{a_m P_n^{(m)}(\alpha)}. \qquad (2.4.90)$$

Example. Find a particular solution of the ODE

$$Ly \equiv y''' - 3y'' + 3y' - y = e^x. \qquad (2.4.91)$$

Solution. The equation (2.4.91) is a non-homogeneous linear equation with constant coefficients. The free term is of the exponential form (2.4.75), with $\alpha = 1$.

The equation can be shaped by means of the differential polynomial

$$Ly \equiv \left(\underbrace{D^3 - 3D^2 + 3D - E}_{P(D)} \right) y = e^x. \qquad (2.4.92)$$

The associated characteristic equation is

$$r^3 - 3r^2 + 3r - 1 = 0. \qquad (2.4.93)$$

Obviously, the left side is a cube, so

$$(r-1)^3 = 0, \qquad (2.4.94)$$

hence 1 is a triple root of the characteristic equation. Consequently, we search for a particular solution of the form

$$Y(x) = ax^3 e^x. \qquad (2.4.95)$$

To differentiate it, we use again formula (2.4.53), with $u = ax^3$, $v = e^{\alpha x}$. Firstly, we compute

$$\begin{aligned} P'(D) &= 3D^2 - 6D + 3E = 3(D-E)^2, \\ P''(D) &= 6D - 6E = 3!(D-E), \\ P'''(D) &= 6E. \end{aligned} \qquad (2.4.96)$$

Obviously,

$$P(D)e^x = 0, \quad P'(D)e^x = 0,$$
$$P''(D)e^x = 0, \quad P'''(D)e^x = 6e^x. \tag{2.4.97}$$

Applying now formula (2.4.53), we obtain

$$P(D)(ax^3 e^x) = ax^3 \underbrace{P(e^x)}_{=0} + 3ax^2 \underbrace{P'(e^x)}_{=0} +$$
$$+ \frac{1}{2!} 6ax \underbrace{P''(e^x)}_{=0} + \frac{1}{3!} 3! a \cdot P'''(e^x) = e^x, \tag{2.4.98}$$

whence, taking into account (2.4.97), we get

$$a = \frac{1}{6}. \tag{2.4.99}$$

Therefore, for the ODE (2.4.91) we found the particular solution

$$\boxed{Y(x) = \frac{1}{6} x^3 e^x}. \tag{2.4.100}$$

C. The free term is an exponential function multiplied by a polynomial:

$$f(x) = P_m(x) e^{\alpha x}. \tag{2.4.101}$$

Again we distinguish two cases:

- **α is not a root of the characteristic equation.** In this case, we search for a particular solution of the non-homogeneous ODE in the form of the free term, i.e.,

$$Y(x) = Q_m(x) e^{\alpha x}. \tag{2.4.102}$$

Example. Find a particular solution of the ODE

$$Ly \equiv y'' - 3y' + 2y = xe^{3x}. \qquad (2.4.103)$$

Solution. The equation (2.4.103) is a non-homogeneous linear equation with constant coefficients. The free term is of the form (2.4.101), where $\alpha = 3$, and $P_m(x) = x$. We have previously shown that this equation can be also written by using the differential polynomial (2.4.79) and we have computed the roots of the characteristic equation (2.4.80), which do not coincide with α. We therefore search for a particular solution of the form

$$Y(x) = (ax + b)e^{3x}. \qquad (2.4.104)$$

To differentiate it, we use formula (2.4.53), for $u = ax + b$, $v = e^{3x}$. Taking into account that

$$P'(D) = 2D - 3E,$$
$$P''(D) = 2E, \qquad (2.4.105)$$

we obtain

$$\begin{aligned} P(D)\big((ax+b)e^{3x}\big) &= \\ &= (ax+b)\cdot P(e^{3x}) + aP'(e^{3x}) = \\ &= (ax+b)(9 - 3\cdot 3 + 2) + a\cdot(6 - 3) = \\ &= (2ax + 3a + 2b)e^{3x}, \end{aligned} \qquad (2.4.106)$$

whence

$$(2ax + 3a + 2b)e^{3x} = xe^{3x} \;\rightarrow\; a = \frac{1}{2},\, b = -\frac{3}{4}. \qquad (2.4.107)$$

Hence, a particular solution of (2.4.103) is

$$\boxed{Y(x) = \frac{1}{4}(2x-3)e^{3x}}. \qquad (2.4.108)$$

- α is ***a multiple root*** of order r, $r \leq n$, ***of the characteristic equation***. In this case, we search for a particular solution Y of the form

$$Y(x) = x^r Q_m(x) e^{\alpha x}. \qquad (2.4.109)$$

Let us note that, in both cases, the formula (2.4.53) is very useful.

Remark. If α is a multiple root of the characteristic equation, then it is easier to use firstly the change of function

$$y(x) = z(x) e^{\alpha x}. \qquad (2.4.110)$$

By applying formula (2.4.53), we obtain an ODE in z, where the exponential is simplified and whose free term is a polynomial; therefore, we are in one of the cases A.

Example. Find a particular solution of the ODE

$$Ly \equiv y''' - 3y'' + 3y' - y = x^5 e^x. \qquad (2.4.111)$$

Solution. The equation (2.4.111) is a non-homogeneous linear equation with constant coefficients. The free term is of the form (2.4.101), with $\alpha = 1$. We have previously written the equation by using the differential polynomial (2.4.92) and we have shown that its characteristic equation allows 1 as a triple root.

Let us make the change of function

$$y(x) = z(x)e^{\alpha x}, \qquad (2.4.112)$$

using the formula (2.4.53) for $u = z(x)$, $v = e^{\alpha x}$ and taking into account the calculus of the formal derivatives of the differential polynomial from (2.4.96). We obtain

$$P(D)(ze^x) = z\underbrace{P(e^x)}_{=0} + z'\underbrace{P'(e^x)}_{=0} + \\ + \frac{1}{2!}z''\underbrace{P''(e^x)}_{=0} + \frac{1}{3!}z''' \cdot P'''(e^x) = x^5 e^x, \qquad (2.4.113)$$

whence, after simplification by e^x, we deduce

$$z''' = x^5. \qquad (2.4.114)$$

This is a third order non-homogeneous linear ODE in z. One of its particular solutions is immediately found by direct integration

$$Z(x) = \frac{1}{6 \cdot 7 \cdot 8} x^8. \qquad (2.4.115)$$

Therefore, a particular solution for the ODE (2.4.111) is

$$\boxed{Y(x) = \frac{1}{6 \cdot 7 \cdot 8} x^8 e^x}. \qquad (2.4.116)$$

D. *The free term is a trigonometric function* (sin, cos)

$$f(x) = a \sin \alpha x + b \cos \alpha x. \qquad (2.4.117)$$

We distinguish, again, two cases:

- **iα *is not a root of the characteristic equation*.** In this case, we shape the particular solution into the free term form:

$$Y(x) = A\cos\alpha x + B\sin\alpha x. \qquad (2.4.118)$$

Example. Find a particular solution of the ODE

$$Ly \equiv y'' - 5y' + 4y = \cos x. \qquad (2.4.119)$$

Solution. The equation (2.4.119) is a linear non-homogeneous equation with constant coefficients. The free term is of the form (2.4.118), with $\alpha = 1$. The equation can be also written in terms of the differential polynomial

$$Ly \equiv \left(\underbrace{D^2 - 5D + 4E}_{P(D)} \right) y = \cos x. \qquad (2.4.120)$$

The associated characteristic equation is

$$r^2 - 5r + 4 = 0, \qquad (2.4.121)$$

with the real roots

$$r_1 = 1, \quad r_2 = 4. \qquad (2.4.122)$$

We search for a particular solution of the form:

$$Y(x) = a\cos x + b\sin x. \qquad (2.4.123)$$

Differentiating and introducing it into the equation gives

$$L(a\cos x + b\sin x) = (-a\cos x - b\sin x) - 5(-a\sin x + b\cos x) + 4(a\cos x + b\sin x) = \cos x. \qquad (2.4.124)$$

Hence, by identifying the coefficients of the trigonometric functions, we infer

$$\begin{cases} 3a - 5b = 1, \\ 5a + 3b = 0, \end{cases} \rightarrow a = \frac{3}{34}, b = -\frac{5}{34}. \qquad (2.4.125)$$

A particular solution of (2.4.119) is then

$$\boxed{Y(x) = \frac{1}{34}(3\cos x - 5\sin x)}. \qquad (2.4.126)$$

- **iα is a multiple root of mth order *of the characteristic equation*.** In this case, a particular solution of the non-homogeneous ODE can be shaped into the form

$$Y(x) = x^m (a\cos x + b\sin x). \qquad (2.4.127)$$

E. *If the free term is a function of the form*

$$f(x) = P_m(x)(a\cos \alpha x + b\sin \alpha x)e^{\beta x}, \qquad (2.4.128)$$

taking into account the roots of the characteristic equation, we could shape again the particular solution into the free term form.

However, it is easier to make firstly the following change

$$y(x) = z(x)e^{\beta x}, \qquad (2.4.129)$$

to use the formula (2.4.53) and, after simplification with $e^{\beta x}$, to find a particular solution of the equation, according to the considerations made at the point **D**.

2.5. EQUATIONS REDUCIBLE TO ODEs WITH CONSTANT COEFFICIENTS

Suppose we must solve an ODE and we could find a convenient change of variable or function that transforms it into a linear one with constant coefficients, which is easily solved. By using the inverse transformation, we also get the solution of the given equation, starting from the solution of the transformed one. We give here several illustrative examples.

1. Consider the equation

$$Ly \equiv (1-x^2)\frac{d^2 y}{dx^2} - x\frac{dy}{dx} + n^2 y = 0. \tag{2.5.1}$$

Let us make the *change of variable*

$$x = \cos t. \tag{2.5.2}$$

We rewrite the equation:

$$
\begin{array}{c|c|c}
\times n^2 & y = y & \times n^2 \\
\times (-x) & \dfrac{dy}{dx} = \dfrac{dy}{dt} \cdot \dfrac{dt}{dx} = -\dfrac{1}{\sin t}\dfrac{dy}{dt} & \times (-\cos t) \\
\times (1-x^2) & \dfrac{d^2 y}{dx^2} = -\dfrac{1}{\sin t}\dfrac{d}{dt}\left(-\dfrac{1}{\sin t}\dfrac{dy}{dt}\right) = & \times \sin^2 t \\
 & = -\dfrac{\cos t}{\sin^3 t}\dfrac{dy}{dt} + \dfrac{1}{\sin^2 t}\dfrac{d^2 y}{dt^2} & \\
\end{array}
$$

$$Ly = \frac{d^2 y}{dt^2} - \frac{\cos t}{\sin t}\frac{dy}{dt} + \frac{\cos t}{\sin t}\frac{dy}{dt} + n^2 y = 0,$$

and it follows that

$$\frac{d^2 y}{dt^2} + n^2 y = 0. \tag{2.5.3}$$

This is a homogeneous linear ordinary differential equation with constant coefficients. According to the previous considerations (see the harmonic oscillator from section 2.4), a fundamental system of solutions for (2.5.3) is

$$y_1 = \cos nt, \quad y_2 = \sin nt, \tag{2.5.4}$$

or, getting back to the variable x,

$$\boxed{y_1 = \cos n(\arccos x), \quad y_2 = \sin n(\arccos x)}. \tag{2.5.5}$$

If $n = 1$, then $y_1 = \cos(\arccos x) = x$. Using this remark, we prove that, for odd n, y_1 is a polynomial of degree n in x.

These polynomials are called ***Chebyshev's polynomials***.

2. Consider ***Bessel's equation***, often used in engineering:

$$x^2 y'' + x y' + \left(x^2 - v^2 \right) y = 0. \tag{2.5.6}$$

This is a second-order linear equation with variable coefficients. Its series form solutions are the so-called ***Bessel's functions***, which depend on the index v.

Let us write Bessel's equation for $v = \dfrac{1}{2}$:

$$Ly \equiv x^2 y'' + x y' + \left(x^2 - \dfrac{1}{4} \right) y = 0. \tag{2.5.7}$$

We apply to this equation the following ***change of function***

$$y = \dfrac{z}{\sqrt{x}}. \tag{2.5.8}$$

Introducing this in (2.5.7), one is led to:

$$
\begin{array}{l|l}
y = x^{-\frac{1}{2}}z & \times\left(x^2 - \dfrac{1}{4}\right) \\[2mm]
y' = x^{-\frac{1}{2}}z' - \dfrac{1}{2}x^{-\frac{3}{2}}z & \times x \\[2mm]
y'' = x^{-\frac{1}{2}}z'' - 2\cdot\dfrac{1}{2}x^{-\frac{3}{2}}z' + \dfrac{3}{4}x^{-\frac{5}{2}}z & \times x^2
\end{array}
$$

$$Ly = x^{\frac{3}{2}}z'' + \left(x^{\frac{1}{2}} - x^{\frac{1}{2}}\right)z' + \left(x^{\frac{3}{2}} - \dfrac{1}{4}x^{-\frac{1}{2}} - \dfrac{1}{2}x^{\frac{1}{2}} + \dfrac{3}{4}x^{-\frac{1}{2}}\right)z = 0,$$

whence we infer for z the following homogeneous linear differential equation with constant coefficients,

$$z'' + z = 0. \tag{2.5.9}$$

The general solution of this equation is

$$z = c_1 \cos x + c_2 \sin x; \tag{2.5.10}$$

getting back to y by (2.5.8), we obtain

$$\boxed{y = \dfrac{c_1 \cos x + c_2 \sin x}{\sqrt{x}}}, \tag{2.5.11}$$

which is **the general solution of Bessel's equation for** $\nu = \dfrac{1}{2}$.

3. EULER'S EQUATION. Consider the nth order linear ordinary differential equation with variable coefficients:

$$a_0 x^n y^{(n)} + a_1 x^{n-1} y^{(n-1)} + \ldots + a_{n-1} x y' + a_n y = 0. \tag{2.5.12}$$

This is *Euler's equation*; we notice that the derivatives of kth order of y are multiplied by x to k^{th} power. We make the change of variable

$$x = e^t. \qquad (2.5.13)$$

For a better understanding, we shall tackle the case $n = 3$; the case of arbitrary n is dealt with analogously.

Therefore, let us consider the ODE

$$a_0 x^3 y''' + a_1 x^2 y'' + a_2 x y' + a_3 y = 0. \qquad (2.5.14)$$

We have

$$\frac{dy}{dx} = \frac{dy}{dt} \cdot \frac{dt}{dx} = e^{-t} \frac{dy}{dt} \quad \rightarrow \quad \boxed{\frac{d}{dx} = e^{-t} \frac{d}{dt}}. \qquad (2.5.15)$$

We rewrite the equation, denoting the identity operator ($Ey = y$) by E:

$\times a_3$	$y = y$	a_3
$\times a_2 x$	$\dfrac{dy}{dx} = e^{-t} \dfrac{dy}{dt}$	$a_2 e^t$
$\times a_1 x^2$	$\dfrac{d^2 y}{dx^2} = e^{-t} \dfrac{d}{dt}\left(e^{-t} \dfrac{dy}{dt}\right) = e^{-2t} \dfrac{d}{dt}\left(\dfrac{d}{dt} - E\right) y$	$a_1 e^{2t}$
$\times a_0 x^3$	$\dfrac{d^3 y}{dx^3} = e^{-t} \dfrac{d}{dt}\left[e^{-2t} \dfrac{d}{dt}\left(\dfrac{dy}{dt} - y\right)\right] =$ $= e^{-3t} \dfrac{d}{dt}\left(\dfrac{d}{dt} - E\right)\left(\dfrac{d}{dt} - 2E\right) y$	$a_0 e^{3t}$

$$Ly = a_0 \frac{d}{dt}\left(\frac{d}{dt} - E\right)\left(\frac{d}{dt} - 2E\right) y + a_1 \frac{d}{dt}\left(\frac{d}{dt} - E\right) y +$$
$$+ a_2 \frac{d}{dt} y + a_3 y = 0.$$

We obtain the linear ODE with constant coefficients

$$Ly = a_0 \frac{d}{dt}\left(\frac{d}{dt} - E\right)\left(\frac{d}{dt} - 2E\right)y +$$
$$+ a_1 \frac{d}{dt}\left(\frac{d}{dt} - E\right)y + a_2 \frac{d}{dt}y + a_3 y = 0. \quad (2.5.16)$$

As we have shown in section 2.4, we search for solutions of the form

$$y = e^{rt}. \quad (2.5.17)$$

The associated characteristic equation is

$$a_0 r(r-1)(r-2) + a_1 r(r-1) + a_2 r + a_3 = 0. \quad (2.5.18)$$

After having solved it, we find a fundamental system of solutions and we write the general solution of the ODE (2.5.16).

Remark. By combining the change of variable with the exponential form (2.5.17), we notice that

$$y = e^{rt} = e^{r \ln x} = e^{\ln x^r} = x^r. \quad (2.5.19)$$

Therefore, in applications, it is easier to search directly for solutions of the form

$$y = x^r. \quad (2.5.20)$$

Example. Find the general solution of this equation

$$x^2 y'' + 3xy' + 5y = 0. \quad (2.5.21)$$

Solution. This is an Euler equation, therefore we search for solutions of the form (2.5.20). We get

$$\begin{array}{c|l} \times 5 & y = x^r \\ \times 3x & y' = rx^{r-1} \\ \times x^2 & y'' = r(r-1)x^{r-2} \end{array} \qquad (2.5.22)$$

$$Ly = \left[r(r-1) + 3r + 5 \right] x^r = 0,$$

therefore, the characteristic equation is

$$r^2 + 2r + 5 = 0, \qquad (2.5.23)$$

with the roots $r_{1,2} = -1 \pm 2i$. The correspondent solutions will be

$$y_1 = x^{-1+2i}, \quad y_2 = x^{-1-2i}. \qquad (2.5.24)$$

In order to keep the real frame, we use Euler's formulas. We have

$$y_1 = x^{-1+2i} = \frac{e^{2i \ln x}}{x}, \quad y_2 = \overline{y_1} = x^{-1-2i} = \frac{e^{-2i \ln x}}{x}, \qquad (2.5.25)$$

so that, using both the real and the imaginary part of y_1, we obtain the real fundamental system of solutions

$$Y_1 = \frac{\cos(2 \ln x)}{x}, \quad Y_2 = \frac{\sin(2 \ln x)}{x}; \qquad (2.5.26)$$

the general solution of Euler's equation (2.5.21) is thus

$$\boxed{y(x) = \frac{c_1 \cos(2 \ln x) + c_2 \sin(2 \ln x)}{x}}. \qquad (2.5.27)$$

EXERCISES AND PROBLEMS

1. Integrate the following high order homogeneous linear differential equations with constant coefficients, also using the conditions (where required):

I. The characteristic equation allows real and distinct roots:

a) $y'' - y' = 0$ \hspace{2em} A: $y = C_1 + C_2 e^x$

b) $y'' - y' - 2y = 0$ \hspace{2em} A: $y = C_1 e^{2x} + C_2 e^{-x}$

c) $y''' - 3y'' - y' + 3y = 0$ \hspace{1em} A: $y = C_1 e^{-x} + C_2 e^x + C_3 e^{3x}$

d) $y'' + 5y' + 6y = 0$
$y(0) = 1, y'(0) = -6$ \hspace{2em} A: $y = 4e^{-3x} - 3e^{-2x}$

e) $y'' + 3y' = 0$
$y(0) = 1, y'(0) = 2$ \hspace{2em} A: $y = \dfrac{1}{3}(5 - 2e^{-3x})$

II. The characteristic equation allows complex roots:

a) $y'' + y' + y = 0$

$$A: y = e^{\frac{-x}{2}}\left(C_1 \cos\frac{\sqrt{3}}{2}x + C_2 \sin\frac{\sqrt{3}}{2}x\right)$$

b) $y'' - 2y' + 10y = 0$
$y\left(\dfrac{\pi}{6}\right) = 0, y'\left(\dfrac{\pi}{6}\right) = e^{\frac{\pi}{6}}$ \hspace{1em} A: $y = -\dfrac{1}{3} e^x \cos 3x$

c) $y^{(4)} + 10y'' + 9y = 0$ $\quad$ A: $y = C_1 \cos x + C_2 \sin x +$
$\qquad\qquad\qquad\qquad\qquad\qquad + C_3 \cos 3x + C_4 \sin 3x$

d) $y'' + 25y = 0$ $\quad$ A: $y = C_1 \cos 5x + C_2 \sin 5x$

e) $y^{(4)} + 5y'' + 4y = 0$ $\quad$ A: $y = C_1 \cos x + C_2 \sin x +$
$\qquad\qquad\qquad\qquad\qquad\qquad + C_3 \cos 2x + C_4 \sin 2x$

**f)
$9y'' + y = 0$
$y\left(\dfrac{3\pi}{2}\right) = 2,\ y'\left(\dfrac{3\pi}{2}\right) = 0$ $\quad$ A: $y = 2\sin\dfrac{x}{3}$

g)
$y'' + 9y = 0$
$y(0) = 0,\ y\left(\dfrac{\pi}{4}\right) = 1$ $\quad$ A: $y = \sqrt{2}\sin 3x$

h)
$y'' + y = 0$
$y'(0) = 1,\ y'\left(\dfrac{\pi}{3}\right) = 0$ $\quad$ A: $y = \sin x + \dfrac{1}{\sqrt{3}}\cos x$

III. The characteristic equation allows multiple roots:

a) $y'' - 4y' + 4y = 0$ $\qquad$ A: $y = (C_1 + C_2 x)e^{2x}$

b)
$y'' - 10y' + 25y = 0$
$y(0) = 0,\ y'(0) = 1$ $\qquad$ A: $y = xe^{5x}$

c) $y''' - 3y'' + 3y' - y = 0$ $\qquad$ A: $y = e^x(C_1 + C_2 x + C_3 x^2)$

d) $y^{(4)} + 2y'' + y = 0$ A: $y = C_1 \cos x + C_2 \sin x + x(C_3 \cos x + C_4 \sin x)$

e) $y^{(4)} - 2y''' + y'' = 0$ A: $y = C_1 + C_2 x + C_3 e^x + C_4 x e^x$

f) $y^{(5)} - 2y^{(4)} - y''' + 2y'' = 0$

A: $y = C_1 + C_2 x + C_3 e^{-x} + C_4 e^x + C_5 e^{2x}$

2. Integrate, also using the conditions (where required), the following high order non-homogeneous linear differential equations with constant coefficients:

I. Using the variation of parameters:

a) $y'' + y = \tan x$

A: $y = C_1 \cos x + C_2 \sin x - \cos x \ln \left| \dfrac{\tan \dfrac{x}{2} - 1}{\tan \dfrac{x}{2} + 1} \right|$

b) $y'' - 2y' + y = \dfrac{e^x}{\sqrt{4 - x^2}}$ A: $y = e^x(C_1 + xC_2) + \sqrt{4 - x^2} e^x + x e^x \arcsin \dfrac{x}{2}$

c) $y'' + 4y = \cot 2x$

A: $y = C_1 \cos 2x + C_2 \sin 2x + \dfrac{1}{4} \sin 2x \ln \tan 2x$

II. Shaping it into the free term form:

a) $y'' + 4y = e^x$ A: $y = C_1 \cos 2x + C_2 \sin 2x + \dfrac{1}{5} e^x$

b) $y'' - 5y' + 6y = 2e^{-x}$ A: $y = C_1 e^{2x} + C_2 e^{3x} + \dfrac{1}{6} e^{-x}$

c) $y'' - 5y' + 6y = 5e^{3x}$ A: $y = C_1 e^x + C_2 e^{2x} +$
$$+ \dfrac{1}{2} e^{3x} \left(x^2 - 2x + 2 \right)$$

d) $y^{(4)} - y = e^x$ A: $y = C_1 e^x + C_2 e^{-x} + C_3 \cos x +$
$$+ C_4 \sin x + \dfrac{1}{4} x e^x$$

e) $y'' - (\alpha + \beta) y' + \alpha \beta y = a e^{\alpha x} + b e^{\beta x}$

A: $y = C_1 e^{\alpha x} + C_2 e^{\beta x} +$
$$+ \dfrac{x}{\alpha - \beta} \left(a e^{\alpha x} - b e^{\beta x} \right)$$

f) $y'' + 3y' - 10y = x e^{-2x}$ A: $y = C_1 e^{2x} + C_2 e^{-5x} +$
$$+ \dfrac{1}{144} (1 - 12x) e^{-2x}$$

g) $y'' - 9y' + 20y = x^2 e^{4x}$ A: $y = C_1 e^{5x} + C_2 e^{4x} -$
$$- \left(\dfrac{1}{3} x^3 + x^2 + 2x \right) e^{4x}$$

h) $y'' - 9y = (x + 2) e^{3x}$ A: $y = C_1 e^{-3x} + C_2 e^{3x} +$
$$+ \dfrac{x}{36} (3x + 11) e^{3x}$$

i) $y'' - 3y' + 2y = (x^2 + x) e^{3x}$ A: $y = C_1 e^x + C_2 e^{2x} +$
$$+ \left(\dfrac{x^2}{2} - x + 1 \right) e^{3x}$$

j) $y'' - 2y' + 2y = e^x \sin x$ 　　$A: y = e^x(C_1 \cos x + C_2 \sin x) - \frac{1}{2} x e^x \cos x$

k) $y'' + 9y = 2\cos 3x + 5\sin 3x$

$$A: y = C_1 \cos 3x + C_2 \sin 3x + \frac{x}{6}(2\sin 3x - 5\cos 3x)$$

l) $\begin{array}{l} y'' + 4y = \cos 2x \\ y(0) = y\left(\frac{\pi}{4}\right) = 0 \end{array}$ 　　$A: y = \frac{1}{16}(4x - \pi)\sin 2x$

m) $\begin{array}{l} y'' + 4y = \sin 2x + 1 \\ y(0) = \frac{1}{4}, y'(0) = 0 \end{array}$ 　　$A: y = \frac{1}{8}\sin 2x - \frac{1}{4}(x\cos 2x - 1)$

n) $\begin{array}{l} y'' + y = \sin x + \cos 2x \\ y(0) = 0, y'(0) = 0 \end{array}$ 　　$A: y = \frac{1}{3}\cos x + \frac{1}{2}\sin x - \frac{x}{2}\cos x - \frac{1}{3}\cos 2x$

o) $y^{(4)} + 2y''' + 5y'' + 8y' + 4y = 40e^x + \cos x$

$$A: y = (C_1 + C_2 x)e^{-x} + C_3 \cos 2x + C_4 \sin 2x + \frac{1}{6}\sin x + 2e^x$$

p) $y'' - 2y' + y = \sin x + e^{-x} + e^x$

$$A: y = (C_1 + C_2 x)e^x + \frac{1}{2}\cos x + \frac{1}{4}e^{-x} + \frac{1}{2}e^x x^2$$

3. Integrate the following Euler's equations, also applying the conditions (where required):

a) $x^2 y'' + x y' + y = 0$ $\quad$ A: $y = C_1 \cos\ln x + C_2 \sin\ln x$

b) $x^2 y'' - x y' + y = 0$ $\quad$ A: $y = (C_1 + C_2 \ln x)x$

c) $x^2 y'' - x y' + 2y = 0$ $\quad$ A: $y = x(C_1 \cos\ln x + C_2 \sin\ln x)$

d) $x^2 y'' - 2x y' + 2y = x$ $\quad$ A: $y = C_1 x + C_2 x^2 - x \ln x$

e) $x^3 y''' + x y' - y = 2x^2$

$$A: y = (C_1 + C_2 \ln x + C_3 \ln^2 x)x + 2x^2$$

f) $x^2 y'' - 3x y' + 3y = 3\ln^2 x$

$$A: y = C_1 x + C_2 x^3 + \frac{1}{9}(9\ln^2 x + 24\ln x + 26)$$

g) $x^2 y'' + x y' + y = \sin(2\ln x)$ $\quad$ A: $y = C_1 \cos\ln x + C_2 \sin\ln x - \frac{1}{3}\sin(2\ln x)$

h) $x^2 y'' + 3xy' + y = \dfrac{1}{x}$ 	A: $y = \dfrac{1}{2x}\left(\ln^2 x + 2\ln x + 2\right)$
$y(1) = 1, y'(1) = 0$

i) $x^2 y'' - 3xy' + 4y = \dfrac{1}{2}x^3$ 	A: $y = \dfrac{1}{2}x^3 - \dfrac{x^2}{\ln 2}\ln x$
$y(1) = \dfrac{1}{2}, y(4) = 0$

4. Model and solve the problem of the linear harmonic oscillator acted upon by a periodic force
   a) with a frequency different from its own
   b) with a frequency equal to its own.

   A: *Mathematical model*: $\ddot{y} + \omega^2 y = A\sin at$

   Solution:

   a) $\omega \neq a \rightarrow$
   $$y(t) = A\sin(\omega t + \delta) + \dfrac{A}{a^2 - \omega^2}\sin at,$$
   $A, \delta \in \Re$ arbitrary

   b) $\omega = a \;\rightarrow\; y(t) = A\sin(\omega t + \delta) - \dfrac{A}{2\omega}t\cos\omega t,$
   $A, \delta \in \Re$ arbitrary

5. Model and solve the problem of the linear harmonic oscillator taking into account the resistance to motion, considering the following cases
   a) free oscillator;
   b) oscillator acted upon by a periodic force.

A: a) *Mathematical model*: $\ddot{y} + 2n\dot{y} + \omega^2 y = 0$

Solution: $y(t) = Ae^{-nt}\sin\left(\sqrt{\omega^2 - n^2}\,t + \delta\right)$,

$A, \delta \in \Re$ arbitrary

b) *Mathematical model*: $\ddot{y} + 2n\dot{y} + \omega^2 y = A\sin at$

Solution:

$$y(t) = Ae^{-nt}\sin(\omega_2 t + \delta) + \frac{A}{\sqrt{\omega_1^2 + 4n^2 a^2}}\sin(at + \varphi),$$

$A, \delta \in \Re$ arbitrary, $\omega_1 = \sqrt{\omega^2 - a^2}$,

$\omega_2 = \sqrt{\omega^2 - n^2}$, $\varphi = -\arctan\dfrac{2n}{\omega^2 - a^2}$.

Chapter 3

SYSTEMS OF LINEAR ODEs WITH CONSTANT COEFFICIENTS

The canonical form of such systems is

$$\begin{cases} y'_1 = a_{11}y_1 + a_{12}y_2 + \ldots + a_{1n}y_n + f_1(x), \\ y'_2 = a_{21}y_1 + a_{22}y_2 + \ldots + a_{2n}y_n + f_2(x), \\ \ldots\ldots\ldots\ldots\ldots\ldots\ldots\ldots\ldots\ldots\ldots\ldots\ldots\ldots\ldots\ldots\ldots \\ y'_n = a_{n1}y_1 + a_{n2}y_2 + \ldots + a_{nn}y_n + f_n(x), \end{cases} \quad (3.1.1)$$

where $f_j \in C^0(\mathrm{I}), \mathrm{I} \subseteq \Re, j = \overline{1,n}, \quad a_{ij} \in \Re, i,j = \overline{1,n}$.

3.1. MATRIX FORM OF THE SYSTEM

If we use the vector of the unknown functions

$$\mathbf{y} = (y_1, \ y_2, \ y_3, \ldots, \ y_n)^T, \quad (3.1.2)$$

we can associate a matrix to the system (3.1.1) as follows

$$\mathbf{A} = \begin{pmatrix} a_{11} & a_{12} & \ldots & a_{1n} \\ a_{21} & a_{22} & \ldots & a_{2n} \\ \ldots & \ldots & \ldots & \ldots \\ a_{n1} & a_{n2} & \ldots & a_{nn} \end{pmatrix}; \quad (3.1.3)$$

writing the free terms as a vector, i.e.,

$$\mathbf{f} = (f_1,\ f_2,\ f_3, \ldots,\ f_n)^T, \tag{3.1.4}$$

and the derivatives too

$$\frac{d\mathbf{y}}{dx} = (y_1',\ y_2',\ y_3', \ldots,\ y_n')^T, \tag{3.1.5}$$

the system (3.1.1) can be written in *matrix form*

$$\frac{d\mathbf{y}}{dx} = \mathbf{A}\mathbf{y} + \mathbf{f}. \tag{3.1.6}$$

3.2. LINEAR HOMOGENEOUS ODSs

For $\mathbf{f} = \mathbf{0}$, we obtain the *associated homogeneous linear system*

$$\frac{d\mathbf{y}}{dx} = \mathbf{A}\mathbf{y}. \tag{3.2.1}$$

In order to solve the system (3.2.1), we search for solutions of the form

$$\mathbf{y} = e^{\alpha x} \cdot \mathbf{C},$$
$$\mathbf{C} = (c_1,\ c_2,\ c_3, \ldots,\ c_n)^T,\quad c_j \in \mathcal{R},\ j = \overline{1, n}. \tag{3.2.2}$$

By differentiating with respect to x, we obtain

$$\frac{d\mathbf{y}}{dx} = \alpha e^{\alpha x} \mathbf{C}. \tag{3.2.3}$$

Replacing this in (3.2.1), it results

$$\alpha e^{\alpha x} \mathbf{C} = e^{\alpha x} \mathbf{A} \mathbf{C}. \tag{3.2.4}$$

We simplify by $e^{\alpha x}$.

Let us denote the identity matrix by **E**:

$$\mathbf{E} = \begin{pmatrix} 1 & 0 & \ldots & 0 \\ 0 & 1 & \ldots & 0 \\ \ldots & \ldots & \ldots & \ldots \\ 0 & 0 & 0 & 1 \end{pmatrix}, \quad \mathbf{E} \in \mathfrak{M}_{n \times n}. \qquad (3.2.5)$$

From (3.2.4), we deduce the condition

$$(\mathbf{A} - \alpha \mathbf{E})\mathbf{C} = \mathbf{0}, \qquad (3.2.6)$$

where **0** is the zero vector with n components.

The condition (3.2.6) is, in fact, a homogeneous linear algebraic system of n equations with n unknowns $c_1, c_2, \ldots, c_n$, written in matrix form.

If the system (3.2.1) allows solutions of the form (3.2.2), then

> α is *the eigenvalue* of the matrix **A** of the system and
>
> **C** is the *correspondent eigenvector*.

Therefore, the problem of solving the homogeneous differential system (3.2.1) was reduced to a

PROBLEM OF ALGEBRA: *find the eigenvalues and the eigenvectors of the associated matrix* (3.2.5).

In order for (3.2.6) to allow non zero solutions, one must have

$$\det(\mathbf{A} - \alpha \mathbf{E}) = 0, \qquad (3.2.7)$$

or, more explicitely,

$$\begin{vmatrix} a_{11} - \alpha & a_{12} & \ldots & a_{1n} \\ a_{21} & a_{22} - \alpha & \ldots & a_{2n} \\ \ldots & \ldots & \ldots & \ldots \\ a_{n1} & a_{n2} & \ldots & a_{nn} - \alpha \end{vmatrix} = 0. \quad (3.2.8)$$

The condition (3.2.7), or its equivalent, (3.2.8), represents a polynomial algebraic equation, called ***the characteristic equation.***

As in the case of ODEs, the solution of the ODS (3.2.1) depends on the roots of its characteristic equation.

For illustration, let us consider the case $n = 3$.

$$\begin{cases} \dfrac{dy_1}{dx} = a_{11}y_1 + a_{12}y_2 + a_{13}y_3, \\ \dfrac{dy_2}{dx} = a_{21}y_1 + a_{22}y_2 + a_{23}y_3, \\ \dfrac{dy_3}{dx} = a_{31}y_1 + a_{32}y_2 + a_{33}y_3. \end{cases} \quad (3.2.9)$$

The matrix associated to this system is

$$\mathbf{A} \equiv \begin{pmatrix} a_{11} & a_{12} & a_{13} \\ a_{21} & a_{22} & a_{23} \\ a_{31} & a_{32} & a_{33} \end{pmatrix}, \quad (3.2.10)$$

and the characteristic equation reads

$$\begin{vmatrix} a_{11} - \alpha & a_{12} & a_{13} \\ a_{21} & a_{22} - \alpha & a_{23} \\ a_{31} & a_{32} & a_{33} - \alpha \end{vmatrix} = 0, \quad (3.2.11)$$

having three roots, $\alpha_1, \alpha_2, \alpha_3$. We distinguish three cases:

A. *The case of the real and distinct roots*

Let $\alpha_1, \alpha_2, \alpha_3 \in \Re$ be distinct. Then we have the following diagram:

$$\begin{matrix} \alpha_1 & \alpha_2 & \alpha_3 \\ \downarrow & \downarrow & \downarrow \\ e^{\alpha_1 x} & e^{\alpha_2 x} & e^{\alpha_3 x} \end{matrix}. \qquad (3.2.12)$$

Let $\mathbf{C}_1, \mathbf{C}_2, \mathbf{C}_3$ be the eigenvectors corresponding respectively to $\alpha_1, \alpha_2, \alpha_3$.

According to the previous considerations, it follows that $\{e^{\alpha_1 x}\mathbf{C}_1, e^{\alpha_2 x}\mathbf{C}_2, e^{\alpha_3 x}\mathbf{C}_3\}$ form a fundamental system of solutions.

Therefore, *the general solution* of the system (3.2.9) is

$$\mathbf{y}(x) = k_1 \mathbf{C}_1 \cdot e^{\alpha_1 x} + k_2 \mathbf{C}_2 \cdot e^{\alpha_2 x} + k_3 \mathbf{C}_3 \cdot e^{\alpha_3 x}, \qquad (3.2.13)$$

where k_1, k_2, k_3 are arbitrary constants.

B. *The case of complex roots*

If the equation allows a complex root, then, as it has real coefficients, it will also allow its complex conjugate as solution.

Therefore, let $\alpha_1 = \alpha + i\beta$, $\alpha_2 = \alpha - i\beta$ be complex conjugate roots and $\alpha_3 \in \Re$.

Let $\mathbf{C} + i\mathbf{D}$ be the eigenvector associated to $\alpha_1 = \alpha + i\beta$. Then $\mathbf{C} - i\mathbf{D}$ is the eigenvector for $\alpha_2 = \alpha - i\beta$ (see e.g. [2][5]). Therefore, we have the diagram

$$\begin{array}{ccc} \alpha+i\beta & \alpha-i\beta & \alpha_3 \\ \downarrow & \downarrow & \downarrow \\ (C+iD)e^{(\alpha+i\beta)x} & (C-iD)e^{(\alpha-i\beta)x} & e^{\alpha_3 x}C_3. \end{array} \qquad (3.2.14)$$

Instead of considering the straightforwardly found complex functions, we shall use some real linear combinations of them.

Let us note that the sum of the first two solutions divided by 2 is a real function y_1, which is also a solution of the system, as the system is linear. We get

$$\begin{aligned} y_1 &= \frac{1}{2}\left[(C+iD)e^{(\alpha+i\beta)x} + (C-iD)e^{(\alpha-i\beta)x}\right] = \\ &= e^{\alpha x}\left[C\frac{e^{i\beta x}+e^{-i\beta x}}{2} + iD\frac{e^{i\beta x}-e^{-i\beta x}}{2}\right] = \\ &= e^{\alpha x}\left[C\underbrace{\frac{e^{i\beta x}+e^{-i\beta x}}{2}}_{\cos\beta x} - D\underbrace{\frac{e^{i\beta x}-e^{-i\beta x}}{2i}}_{\sin\beta x}\right], \end{aligned} \qquad (3.2.15)$$

whence

$$y_1 = e^{\alpha x}(C\cos\beta x - D\sin\beta x). \qquad (3.2.16)$$

This is, in fact, ***the real part*** of $(C+iD)e^{(\alpha+i\beta)x}$.

Subtracting the first two solutions from (3.2.14) and dividing the difference by 2i, we obtain another real solution y_2:

$$y_2 = \frac{1}{2i}\left[(C+iD)e^{(\alpha+i\beta)x} - (C-iD)e^{(\alpha-i\beta)x}\right] =$$

$$= e^{\alpha x}\left[C\frac{e^{i\beta x} - e^{-i\beta x}}{2i} + iD\frac{e^{i\beta x} + e^{-i\beta x}}{2i}\right] = \qquad (3.2.17)$$

$$= e^{\alpha x}\left[C\underbrace{\frac{e^{i\beta x} - e^{-i\beta x}}{2i}}_{\sin\beta x} + D\underbrace{\frac{e^{i\beta x} + e^{-i\beta x}}{2}}_{\cos\beta x}\right],$$

therefore

$$y_2 = e^{\alpha x}(C\sin\beta x + D\cos\beta x), \qquad (3.2.18)$$

which is, in fact, ***the imaginary part*** of $(C+iD)e^{(\alpha+i\beta)x}$.

Finally, the general solution of the system (3.2.9) is, in this case,

$$\begin{aligned}y &= k_1 e^{\alpha x}(C\cos\beta x - D\sin\beta x) + \\ &+ k_2 e^{\alpha x}(C\sin\beta x + D\cos\beta x) + k_3 e^{\alpha_3 x}C_3.\end{aligned} \qquad (3.2.19)$$

with k_1, k_2, k_3 arbitrary constants.

C. *The case of multiple roots*

If the equation allows a double root $\alpha_1 = \alpha_2 = \alpha$, with the third one being different from it, i.e., $\alpha_3 \neq \alpha$, then the diagram of the fundamental system

$$\begin{array}{ccc} \alpha & \alpha & \alpha_3 \\ \downarrow & \downarrow & \downarrow \\ e^{\alpha x}C & ? & e^{\alpha_3 x}C_3 \end{array} \qquad (3.2.20)$$

is, so far, incomplete.

In the above diagram, $\mathbf{C}$ and $\mathbf{C}_3$ are the eigenvectors correspondent to α, respectively α_3. Clearly, we must find a different solution for the fundamental system. As in the similar case of linear ODE, whose characteristic equation allows multiple roots, we search for solutions of the form

$$\mathbf{y} = x\mathbf{C}e^{\alpha x} + \mathbf{D}e^{\alpha x}. \qquad (3.2.21)$$

Replacing it in (3.2.9), we find $\mathbf{D}$ from the algebraic system, which is obtained after simplifying by $e^{\alpha x}$. The general solution of the system (3.2.9), is thus

$$\mathbf{y} = k_1\mathbf{C}e^{\alpha x} + k_2(\mathbf{C}x + \mathbf{D})e^{\alpha x} + k_3\mathbf{C}_3 e^{\alpha_3 x}. \qquad (3.2.22)$$

Here, k_1, k_2, k_3 are arbitrary constants.

Examples

1. **Case A).** Solve the system

$$\begin{cases} \dfrac{dy}{dx} = z, \\ \dfrac{dz}{dx} = y. \end{cases} \qquad (3.2.23)$$

Solution. This is a linear and homogeneous system of differential equations. The matrix associated to the system is

$$\mathbf{A} = \begin{pmatrix} 0 & 1 \\ 1 & 0 \end{pmatrix}. \qquad (3.2.24)$$

Denoting the vector of the unknown functions by $\mathbf{y} = \begin{pmatrix} y \\ z \end{pmatrix}$ and the vector of their derivatives by $\mathbf{y}' = \begin{pmatrix} y' \\ z' \end{pmatrix}$, we write the system in matrix form

$$\mathbf{y}' = \mathbf{A}\mathbf{y}.$$

Searching for solutions of the form $e^{\alpha x}\mathbf{C}$, we are led to the characteristic equation

$$\det(\mathbf{A} - \alpha \mathbf{E}) \equiv \begin{vmatrix} -\alpha & 1 \\ 1 & -\alpha \end{vmatrix} = \alpha^2 - 1 = 0,$$

whose solutions are real and distinct: $\alpha_1 = 1, \alpha_2 = -1$.

The eigenvectors are

$$\mathbf{C}_1 = \begin{pmatrix} 1 \\ 1 \end{pmatrix}, \quad \mathbf{C}_2 = \begin{pmatrix} 1 \\ -1 \end{pmatrix}. \tag{3.2.25}$$

Consequently,

$$\mathbf{y}_1 = \begin{pmatrix} 1 \\ 1 \end{pmatrix} e^x, \quad \mathbf{y}_2 = \begin{pmatrix} 1 \\ -1 \end{pmatrix} e^{-x} \tag{3.2.26}$$

form a fundamental system of solutions for the system (3.2.23). Its general solution is

$$\mathbf{y}(x) = k_1 \begin{pmatrix} 1 \\ 1 \end{pmatrix} e^x + k_2 \begin{pmatrix} 1 \\ -1 \end{pmatrix} e^{-x}, \tag{3.2.27}$$

or, by components

$$\begin{cases} y(x) = k_1 e^x + k_2 e^{-x} \\ z(x) = k_1 e^x - k_2 e^{-x} \end{cases}, \quad k_1, k_2 \in \Re, \text{ arbitrary.} \qquad (3.2.28)$$

2. Case B). Solve the system

$$\begin{cases} \dfrac{dy}{dx} = z, \\ \dfrac{dz}{dx} = -y. \end{cases} \qquad (3.2.29)$$

Solution. The associated of the system is

$$\mathbf{A} = \begin{pmatrix} 0 & 1 \\ -1 & 0 \end{pmatrix}, \qquad (3.2.30)$$

and, using the same notations as in the previous exercise, we obtain the matrix form of the system:

$$\mathbf{y}' = \mathbf{A}\mathbf{y}. \qquad (3.2.31)$$

The characteristic equation

$$\det(A - \alpha E) \equiv \begin{vmatrix} -\alpha & 1 \\ -1 & -\alpha \end{vmatrix} = \alpha^2 + 1 = 0 \qquad (3.2.32)$$

has the pure imaginary roots $\alpha_1 = i, \quad \alpha_2 = -i$.

The correspondent eigenvectors are complex conjugate

$$\mathbf{C} = \begin{pmatrix} 1 \\ i \end{pmatrix}, \quad \overline{\mathbf{C}} = \begin{pmatrix} 1 \\ -i \end{pmatrix}. \qquad (3.2.33)$$

We obtain the fundamental system

$$\mathbf{Y} = \begin{pmatrix} 1 \\ i \end{pmatrix} e^{ix}, \quad \overline{\mathbf{Y}} = \begin{pmatrix} 1 \\ -i \end{pmatrix} e^{-ix}. \qquad (3.2.34)$$

We divide by 2 their sum and their difference by 2i, thus getting

$$y_1 = \operatorname{Re} \mathbf{Y} = \frac{\mathbf{Y}+\bar{\mathbf{Y}}}{2} = \frac{1}{2}\begin{pmatrix} e^{ix}+e^{-ix} \\ i(e^{ix}-e^{-ix}) \end{pmatrix} = \begin{pmatrix} \cos x \\ -\sin x \end{pmatrix},$$

$$y_2 = \operatorname{Im} \mathbf{Y} = \frac{\mathbf{Y}-\bar{\mathbf{Y}}}{2i} = \begin{pmatrix} \dfrac{e^{ix}-e^{-ix}}{2i} \\ \dfrac{e^{ix}+e^{-ix}}{2} \end{pmatrix} = \begin{pmatrix} \sin x \\ \cos x \end{pmatrix}. \quad (3.2.35)$$

Therefore, the general solution of the system (3.2.29) is

$$\mathbf{y} = k_1 \begin{pmatrix} \cos x \\ -\sin x \end{pmatrix} + k_2 \begin{pmatrix} \sin x \\ \cos x \end{pmatrix}, \quad (3.2.36)$$

or, by components,

$$\begin{cases} y(x) = k_1 \cos x + k_2 \sin x \\ z(x) = -k_1 \sin x + k_2 \cos x \end{cases}, \quad k_1, k_2 \in \mathfrak{R}, \text{ arbitrary.} \quad (3.2.37)$$

3. Case C). Solve the system

$$\begin{cases} \dfrac{dy}{dx} = ay + z, \\ \dfrac{dz}{dx} = az. \end{cases} \quad (3.2.38)$$

Solution. Using the associated matrix

$$\mathbf{A} = \begin{pmatrix} a & 1 \\ 0 & a \end{pmatrix}, \quad (3.2.39)$$

and the same notations for the unknown functions and their derivatives as previously, we write the system in matrix form:

$$\mathbf{y}' = \mathbf{A}\mathbf{y}. \qquad (3.2.40)$$

The characteristic equation is

$$\det(\mathbf{A} - \alpha\mathbf{E}) \equiv \begin{vmatrix} a-\alpha & 1 \\ 0 & a-\alpha \end{vmatrix} = (a-\alpha)^2 = 0, \qquad (3.2.41)$$

which has the double root $\alpha_1 = \alpha_2 = a$. The eigenvector correspondng to a is

$$\mathbf{C} = \begin{pmatrix} 1 \\ 0 \end{pmatrix}, \qquad (3.2.42)$$

whence we obtain a first solution of the required fundamental system

$$\mathbf{y}_1 = \begin{pmatrix} 1 \\ 0 \end{pmatrix} e^{ax}. \qquad (3.2.43)$$

We search for $\mathbf{y}_2$ of the form

$$\mathbf{y}_2 = \begin{pmatrix} 1 \\ 0 \end{pmatrix} e^{ax} \cdot x + \begin{pmatrix} d_1 \\ d_2 \end{pmatrix} e^{ax}. \qquad (3.2.44)$$

Replacing it in the system, it is found

$$\begin{pmatrix} 1 \\ 0 \end{pmatrix} e^{ax} + ax \begin{pmatrix} 1 \\ 0 \end{pmatrix} e^{ax} + a \begin{pmatrix} d_1 \\ d_2 \end{pmatrix} e^{ax} =$$
$$= \begin{pmatrix} a & 1 \\ 0 & a \end{pmatrix} \begin{pmatrix} 1 \\ 0 \end{pmatrix} x e^{ax} + \begin{pmatrix} a & 1 \\ 0 & a \end{pmatrix} \begin{pmatrix} d_1 \\ d_2 \end{pmatrix} e^{ax}. \qquad (3.2.45)$$

We now simplify by e^{ax} and we get

$$\begin{pmatrix} 1 \\ 0 \end{pmatrix} + \begin{pmatrix} a d_1 \\ a d_2 \end{pmatrix} = \begin{pmatrix} a d_1 + d_2 \\ a d_2 \end{pmatrix}, \qquad (3.2.46)$$

whence it results that $d_2 = 1$, d_1 being arbitrary. We can take $d_1 = 0$. Hence, the second solution of the fundamental system is

$$\mathbf{y}_2 = \begin{pmatrix} 1 \\ 0 \end{pmatrix} e^{ax} \cdot x + \begin{pmatrix} 0 \\ 1 \end{pmatrix} e^{ax}, \qquad (3.2.47)$$

or

$$\mathbf{y}_2 = \begin{pmatrix} x \\ 1 \end{pmatrix} e^{ax}. \qquad (3.2.48)$$

The general solution of the system is thus

$$\mathbf{y} = k_1 \begin{pmatrix} 1 \\ 0 \end{pmatrix} e^{ax} + k_2 \begin{pmatrix} x \\ 1 \end{pmatrix} e^{ax}, \qquad (3.2.49)$$

or, by components,

$$\begin{cases} y(x) = (k_1 + k_2 x) e^{ax} \\ z(x) = k_2 e^{ax} \end{cases}, \quad k_1, k_2 \in \Re, \text{ arbitrary.} \qquad (3.2.50)$$

Remarks

1. The arbitrary constants of the general solution of the ODS are deduced from supplementary conditions, according to the physical phenomenon modeled by them. For instance, in problems of motion, these conditions are often the so called **Cauchy** or **initial conditions**:

$$y_1(x_0) = y_{10}, \; y_2(x_0) = y_{20}, \ldots, \; y_n(x_0) = y_{n0}, \qquad (3.2.51)$$

or, in other words,

$$\mathbf{y}(x_0) = \mathbf{y}_0, \qquad (3.2.52)$$

where $\mathbf{y}_0 = (y_{10}, y_{20}, \ldots, y_{n0})^T$.

2. The second equation of the system is actually a linear equation in z that can be solved independently of y. For more insight into this problem, see section 3.4.

3.3. LINEAR NON-HOMOGENEOUS SYSTEMS

As in the case of the linear ODEs, one can prove that the general solution of the non-homogeneous system (3.1.6) can be written as follows:

$$\mathbf{y} = \mathbf{Y} + \mathbf{y}_{homogeneous}, \qquad (3.3.1)$$

where $\mathbf{Y}$ is a particular solution of (3.1.6) and $\mathbf{y}_{homogeneous}$ – the general solution of the associated homogeneous system. We search for the particular solution $\mathbf{Y}$ either by using *the variation of parameters*, or shaping it into the free term form, exactly as in the case of linear ODEs with constant coefficients.

Example. Solve the system

$$\begin{cases} \dfrac{dy}{dx} = z + e^{2x}, \\ \dfrac{dz}{dx} = y. \end{cases} \qquad (3.3.2)$$

Solution. The homogeneous associated system is, actually, (3.2.23), whose general solution was previously found. Therefore

$$\mathbf{y}_{homogeneous} = k_1 \begin{pmatrix} 1 \\ 1 \end{pmatrix} e^x + k_2 \begin{pmatrix} 1 \\ -1 \end{pmatrix} e^{-x}. \qquad (3.3.3)$$

We search for a particular solution of the non-homogeneous system of the form

$$\mathbf{Y} = \begin{pmatrix} a \\ b \end{pmatrix} e^{2x}. \tag{3.3.4}$$

Introducing it into the system, it results

$$2 \begin{pmatrix} a \\ b \end{pmatrix} e^{2x} = \begin{pmatrix} 0 & 1 \\ 1 & 0 \end{pmatrix} \begin{pmatrix} a \\ b \end{pmatrix} e^{2x} + \begin{pmatrix} e^{2x} \\ 0 \end{pmatrix}. \tag{3.3.5}$$

Simplifying by e^{2x}, we get the algebraic system in a, b

$$\begin{aligned} 2a &= b + 1, \\ 2b &= a, \end{aligned} \tag{3.3.6}$$

therefore $a = \dfrac{2}{3}, b = \dfrac{1}{3}$.

It follows that

$$\mathbf{Y} = \frac{1}{3} \begin{pmatrix} 2 \\ 1 \end{pmatrix} e^{2x}, \tag{3.3.7}$$

so that the general solution of the non-homogeneous system (3.3.2) is

$$\mathbf{y} = k_1 \begin{pmatrix} 1 \\ 1 \end{pmatrix} e^{x} + k_2 \begin{pmatrix} 1 \\ -1 \end{pmatrix} e^{-x} + \frac{1}{3} \begin{pmatrix} 2 \\ 1 \end{pmatrix} e^{2x}, \tag{3.3.8}$$

or, by components,

$$\begin{aligned} y &= k_1 e^{x} + k_2 e^{-x} + \frac{2}{3} e^{2x}, \\ z &= k_1 e^{x} - k_2 e^{-x} + \frac{1}{3} e^{2x}. \end{aligned} \tag{3.3.9}$$

3.4. THE LINK BETWEEN ODEs AND ODSs

There is an obvious connection between the ODEs and ODSs. Indeed, any ODE, linear or not, which has been developed with respect to the higher derivative of the unknown function, can be brought to the form of a first order ODS. Consider the equation

$$y^{(n)} = f\left(x, y', y'', \ldots, y^{(n-1)}\right), \qquad (3.4.1)$$

where f is continuous with respect to its arguments. Let us put

$$y = y_1, \quad y' = y_2, \quad y'' = y_3, \ldots, y^{(n-1)} = y_n. \qquad (3.4.2)$$

Then

$$\begin{aligned} y_1' &= y_2, \\ y_2' &= y_3, \\ &\ldots\ldots\ldots\ldots \\ y_{n-1}' &= y_n, \\ y_n' &= f(x, y_1, y_2, \ldots, y_n); \end{aligned} \qquad (3.4.3)$$

this is a first order ODS, with n equations and n unknown functions.

Conversely, to a first order ODS of n equations written in canonical form, one can generally associate a unique ODE of order n. The proof is more complicate in the general case, but in the case of linear ODSs with constant coefficients it is easier.

For a beter understanding, let us take the case $n = 3$. We do not take into account the trivial cases, e.g., those in which the system can be, from the very beginning, split in one 2x2 ODS plus one first order ODE, or even in three first order ODEs.

Consider therefore the linear system

$$\begin{cases} y_1' = a_{11}y_1 + a_{12}y_2 + a_{13}y_3 + f_1(x), \\ y_2' = a_{21}y_1 + a_{22}y_2 + a_{23}y_3 + f_2(x), \\ y_3' = a_{31}y_1 + a_{32}y_2 + a_{33}y_3 + f_3(x), \end{cases} \quad (3.4.4)$$

and suppose that f_j are at least of class C^2 on some real interval I. We differentiate now the first equation:

$$y_1'' = a_{11}y_1' + a_{12}y_2' + a_{13}y_3' + f_1'(x), \quad (3.4.5)$$

and we replace y_2', y_3' by the right members from the system, thus getting

$$y_1'' = \underbrace{\left(a_{11}^2 + a_{12}a_{21} + a_{13}a_{31}\right)}_{b_{11}} y_1 + \\ + \underbrace{\left(a_{11}a_{12} + a_{12}a_{22} + a_{13}a_{32}\right)}_{b_{12}} y_2 + \\ + \underbrace{\left(a_{11}a_{13} + a_{12}a_{23} + a_{13}a_{33}\right)}_{b_{13}} y_3 + \\ + \underbrace{f_1'(x) + a_{11}f_1 + a_{12}f_2 + a_{13}f_3}_{F_1}, \quad (3.4.6)$$

which means

$$y_1'' = b_{11}y_1 + b_{12}y_2 + b_{13}y_3 + F_1(x). \quad (3.4.7)$$

We differentiate again (3.4.7), applying the same procedure as before. We thus get a similar expression for y_1''':

$$y_1''' = c_{11}y_1 + c_{12}y_2 + c_{13}y_3 + F_2(x). \quad (3.4.8)$$

The first equation of (3.4.4) and the equation (3.4.7) form an algebraic system in y_2, y_3:

$$a_{12}y_2 + a_{13}y_3 = y_1' - a_{11}y_1 - f_1(x),$$
$$b_{12}y_2 + b_{13}y_3 = y_1'' - b_{11}y_1 - F_1(x). \qquad (3.4.9)$$

a) If the associate determinant $a_{12}b_{13} - a_{13}b_{12}$ does not vanish, then we can obtain the expressions of y_2, y_3 in terms of y_1, y_1', y_1''

$$y_2 = A_1 y_1 + A_2 y_1' + A_3 y_1'' + F_2(x),$$
$$y_3 = B_1 y_1 + B_2 y_1' + B_3 y_1'' + F_3(x). \qquad (3.4.10)$$

Now we replace these expressions in (3.4.8), finally obtaining a linear third order ODE with constant coefficients.

This procedure is easily generalized to arbitrary values of n. Naturally, there are particular cases in which this method can be simplified.

Example

Consider the linear homogeneous ODS with constant coefficients

$$\begin{cases} y_1' = -y_1 + y_2 + y_3, \\ y_2' = y_1 - y_2 + y_3, \\ y_3' = y_1 + y_2 + y_3. \end{cases} \qquad (3.4.11)$$

1. Let us solve it by using the matrix method. The associated matrix is

$$\mathbf{A} = \begin{bmatrix} -1 & 1 & 1 \\ 1 & -1 & 1 \\ 1 & 1 & 1 \end{bmatrix}. \qquad (3.4.12)$$

The characteristic equation is

$$\det(\mathbf{A} - \alpha\mathbf{I}) \equiv \begin{pmatrix} -1-\alpha & 1 & 1 \\ 1 & -1-\alpha & 1 \\ 1 & 1 & 1-\alpha \end{pmatrix} = 0, \qquad (3.4.13)$$

and it has three real and distinct roots: $\alpha_1 = -1,\ \alpha_2 = -2,\ \alpha_3 = 2$. We compute the corresponding eigen vectors; they are

$$\mathbf{C}_1 = \begin{bmatrix} 1 \\ 1 \\ -1 \end{bmatrix},\ \mathbf{C}_2 = \begin{bmatrix} 1 \\ -1 \\ 0 \end{bmatrix},\ \mathbf{C}_3 = \begin{bmatrix} 1 \\ 1 \\ 2 \end{bmatrix}.$$

The general solution of the system, written by components, is thus

$$\begin{aligned} y_1 &= c_1 e^{-x} + c_2 e^{-2x} + c_3 e^{2x}, \\ y_2 &= c_1 e^{-x} - c_2 e^{-x} + c_3 e^{2x}, \\ y_3 &= -c_1 e^{-x} + 2 c_3 e^{2x}. \end{aligned} \qquad (3.4.14)$$

2. Let us use the method of reduction to a third order ODE.

We differentiate the first equation of (3.4.11), replacing y_1', y_2', y_3' from the system. We get

$$\begin{aligned} y_1'' &= -y_1' + y_2' + y_3' = -(-y_1 + y_2 + y_3) + \\ &\quad + y_1 - y_2 + y_3 + y_1 + y_2 + y_3, \end{aligned} \qquad (3.4.15)$$

hence

$$y_1'' = 3y_1 - y_2 + y_3. \tag{3.4.16}$$

We differentiate once more, repeating the procedure. We deduce

$$y_1''' = -3y_1 + 5y_2 + 3y_3. \tag{3.4.17}$$

Now, from the first equation (3.4.11) and from (3.4.16) we get the algebraic system in y_2, y_3:

$$\begin{aligned} y_2 + y_3 &= y_1' + y_1, \\ -y_2 + y_3 &= y_1'' - 3y_1, \end{aligned} \tag{3.4.18}$$

which yields

$$\begin{aligned} 2y_2 &= -y_1'' + y_1' + 4y_1, \\ 2y_3 &= y_1'' + y_1' - 2y_1, \end{aligned} \tag{3.4.19}$$

Replacing this in (3.4.17) gives

$$y_1''' + y_1'' - 4y_1' - 4y_1 = 0, \tag{3.4.20}$$

which is the third order linear homogeneous ODE with constant coefficients we are looking for. The associated characteristic equation is

$$\alpha^3 + \alpha^2 - 4\alpha - 4 = 0, \tag{3.4.21}$$

with the same real and distinct roots as previously found for the system: $\alpha_1 = -1, \alpha_2 = -2, \alpha_3 = 2$. Then the general solution of (3.4.20) is

$$y_1 = c_1 e^{-x} + c_2 e^{-2x} + c_3 e^{2x}. \tag{3.4.22}$$

Taking into account (3.4.19), we also obtain

$$y_2 = c_1 e^{-x} - c_2 e^{-x} + c_3 e^{2x},$$
$$y_3 = -c_1 e^{-x} + 2c_3 e^{2x}. \tag{3.4.23}$$

Looking at (3.4.22) and (3.4.23), it is easily seen that we got the same general solution for the system.

b) If $a_{12}b_{13} - a_{13}b_{12} = 0$, then we may perhaps apply the previous procedure taking y_2 or y_3 into account instead of y_1.

Example

Consider the system

$$\begin{cases} y_1' = y_1 + y_2 + 2y_3, \\ y_2' = 2y_1 + 2y_2 + 4y_3, \\ y_3' = -y_1 + 3y_2 + 6y_3. \end{cases} \tag{3.4.24}$$

Let us solve it by using the matrix method. The associated matrix is

$$\mathbf{A} = \begin{bmatrix} 1 & 1 & 2 \\ 2 & 2 & 4 \\ -1 & 3 & 6 \end{bmatrix}. \tag{3.4.25}$$

The characteristic equation is

$$\det(\mathbf{A} - \alpha \mathbf{I}) \equiv \begin{pmatrix} 1-\alpha & 1 & 2 \\ 2 & 2-\alpha & 4 \\ -1 & 3 & 6-\alpha \end{pmatrix} = 0, \tag{3.4.26}$$

and it has three real and distinct roots: $\alpha_1 = 0$, $\alpha_2 = 1$, $\alpha_3 = 8$. We compute the corresponding eigen vectors; they are

$$\mathbf{C}_1 = \begin{bmatrix} 0 \\ 2 \\ -1 \end{bmatrix}, \mathbf{C}_2 = \begin{bmatrix} 1 \\ 2 \\ -1 \end{bmatrix}, \mathbf{C}_3 = \begin{bmatrix} 1 \\ 2 \\ \dfrac{5}{2} \end{bmatrix}.$$

The general solution of the system, written by components, is thus

$$\begin{aligned} y_1 &= c_2 e^x + c_3 e^{8x}, \\ y_2 &= 2c_1 + 2c_2 e^x + 2c_3 e^{8x}, \\ y_3 &= -c_1 - c_2 e^x + \frac{5}{2} c_3 e^{8x}. \end{aligned} \qquad (3.4.27)$$

Let us use now the method of reduction to a third order ODE. We differentiate the first equation of (3.4.11), replacing y_1', y_2', y_3' from the system. We get

$$\begin{aligned} y_1'' &= y_1' + y_2' + 2y_3' = y_1 + y_2 + 2y_3 + \\ &+ 2y_1 + 2y_2 + 4y_3 + 2(-y_1 + 3y_2 + 6y_3), \end{aligned} \qquad (3.4.28)$$

hence

$$y_1'' = y_1 + 9y_2 + 18 y_3. \qquad (3.4.29)$$

Here, $a_{12}b_{13} - a_{13}b_{12} = 1 \cdot 18 - 2 \cdot 9 = 0$.

Let us apply the above described procedure to y_2.

We differentiate the second equation of (3.4.24), replacing y_1', y_2', y_3' from the system. It results

$$\begin{aligned} y_2'' &= 2y_1' + 2y_2' + 4y_3' = 2(y_1 + y_2 + 2y_3) + \\ &+ 2(2y_1 + 2y_2 + 4y_3) - y_1 + 3y_2 + 6y_3, \end{aligned} \qquad (3.4.30)$$

hence

$$y_2'' = 2y_1 + 18y_2 + 36y_3. \quad (3.4.31)$$

We try now to get y_1, y_3 from the algebraic system formed by the second equation of (3.4.24) and (3.4.31), i.e.

$$\begin{aligned} 2y_1 + 4y_3 &= y_2' - 2y_2, \\ 2y_1 + 36y_3 &= y_2'' - 18y_2, \end{aligned} \quad (3.4.32)$$

We see that the determinant $\begin{vmatrix} 2 & 4 \\ 2 & 18 \end{vmatrix}$ is not null. We immediately obtain

$$\begin{aligned} 2y_1 &= y_2'' - 2y_2' - \frac{y_2'' - y_2' - 16y_2}{8}, \\ 32y_3 &= y_2'' - y_2' - 16y_2. \end{aligned} \quad (3.4.33)$$

Differetiating once (3.4.31) and replacing the first derivatives from the system yields

$$y_2''' = 2y_1 + 2 \cdot 73 y_2 + 4 \cdot 73 y_3, \quad (3.4.34)$$

or, taking (3.4.33) into account,

$$y_2''' - 9y_2'' + 8y_2' = 0, \quad (3.4.35)$$

which is a linear third order ODE, equivalent to the system. The associated characteristic equation is, obviously

$$\alpha^3 - 9\alpha^2 + 8\alpha = 0, \quad (3.4.36)$$

whith the same real and distinct roots as previously found for the system: $\alpha_1 = 0, \alpha_2 = 1, \alpha_3 = 8$. Hence the general solution of (3.4.35) is

$$y_2 = 2c_1 + 2c_2 e^x + 2c_3 e^{8x}, \quad (3.4.37)$$

with c_1, c_2, c_3 arbitrary constants.

Taking into account (3.4.33), we also obtain

$$y_1 = c_2 e^x + c_3 e^{8x},$$
$$y_3 = -c_1 - c_2 e^x + \frac{5}{2} c_3 e^{8x}. \tag{3.4.38}$$

Looking at (3.4.37) and (3.4.38), it easily seen that we got the same general solution for the system.

Remark. We denoted the arbitrary constants such that there be a perfect match with the general solution found by the matrix method. This can be always done.

c) Let us see what happens if the reduction procedure fails for all y_1, y_2, y_3. This means that one has simultaneously

$$\begin{vmatrix} a_{12} & a_{13} \\ b_{12} & b_{13} \end{vmatrix} = 0, \quad \begin{vmatrix} a_{21} & a_{23} \\ b_{21} & b_{23} \end{vmatrix} = 0, \quad \begin{vmatrix} a_{31} & a_{32} \\ b_{31} & b_{32} \end{vmatrix} = 0. \tag{3.4.39}$$

After replacing the coefficients b_{ij} by their expressions as in (3.4.6), an easy calculation gives

$$a_{12}\Delta_{a_{31}} - a_{13}\Delta_{a_{21}} = 0,$$
$$a_{21}\Delta_{a_{32}} - a_{23}\Delta_{a_{12}} = 0, \tag{3.4.40}$$
$$a_{31}\Delta_{a_{23}} - a_{32}\Delta_{a_{13}} = 0,$$

where $\Delta_{a_{ij}}$ signifies the algebraic complement corresponding to a_{ij} in the matrix (3.2.10), corresponding to the system (3.4.4).

This happens, for instance, if the rank of **A** is 1, case in which all the minors of A are null.

Actually, one can proof that if $\Delta_{a_{31}}, \Delta_{a_{32}}, \Delta_{a_{21}}, \Delta_{a_{23}}$ are null, then **A** is of the form

$$\mathbf{A} \equiv \begin{pmatrix} a_{11} & a_{12} & a_{13} \\ ka_{11} & ka_{12} & ka_{13} \\ ma_{11} & ma_{12} & ma_{13} \end{pmatrix}. \tag{3.4.41}$$

Example

Let us consider the system

$$\begin{cases} y_1' = a_{11}y_1 + a_{12}y_2 + a_{13}y_3 \,, \\ y_2' = k(a_{11}y_1 + a_{12}y_2 + a_{13}y_3), \\ y_3' = m(a_{11}y_1 + a_{12}y_2 + a_{13}y_3), \end{cases} \tag{3.4.42}$$

whose associated matrix is precisely (3.4.41). This system can be immediately reduced to a first order equation containing two arbitrary constants. Indeed, we obviously have

$$\begin{cases} ky_1' - y_2' = 0, \\ my_1' - y_3' = 0, \end{cases} \tag{3.4.43}$$

which involves, by direct integration

$$\begin{cases} y_2 = ky_1 + c_2 \,, \\ y_3 = my_1 + c_3 \,, \end{cases} \tag{3.4.44}$$

with c_1, c_2 arbitrary constants. Replacing these expressions in the first equation of the system, we get

$$y_1' = a_{11}y_1 + a_{12}(ky_1 + c_2) + a_{13}(my_1 + c_3), \tag{3.4.45}$$

or

$$y_1' = (a_{11} + ka_{12} + ma_{13}) y_1 + a_{12}c_2 + a_{13}c_3, \qquad (3.4.46)$$

i.e. a first order linear and non-homogeneous equation with constant coefficients, that can be solved by using the method exposed at section 1.3.6. The associated homogeneous equation

$$y_1' = (a_{11} + ka_{12} + ma_{13}) y_1, \qquad (3.4.47)$$

allows the general solution

$$y_{1 homogeneous} = c_1 e^{(a_{11}+ka_{12}+ma_{13})x}, \qquad (3.4.48)$$

and a particular solution of (3.4.46) is obviously given by

$$y_{1 particular} = -\frac{a_{12}c_2 + a_{13}c_3}{a_{11} + ka_{12} + ma_{13}}. \qquad (3.4.49)$$

The general solution of (3.4.46) is therefore

$$y_1 = c_1 e^{(a_{11}+ka_{12}+ma_{13})x} - \frac{a_{12}c_2 + a_{13}c_3}{a_{11} + ka_{12} + ma_{13}}. \qquad (3.4.50)$$

From (3.4.44) we obtain

$$\begin{aligned} y_2 &= kc_1 e^{(a_{11}+ka_{12}+ma_{13})x} + c_2 - k\frac{a_{12}c_2 + a_{13}c_3}{a_{11} + ka_{12} + ma_{13}}, \\ y_3 &= mc_1 e^{(a_{11}+ka_{12}+ma_{13})x} + c_3 - m\frac{a_{12}c_2 + a_{13}c_3}{a_{11} + ka_{12} + ma_{13}}. \end{aligned} \qquad (3.4.51)$$

The general solution of the system (3.4.46) is given by the above expressions (3.4.50) and (3.4.51).

The case $n = 2$ is simpler !

Indeed, let us consider the general first order ODS in two unknown functions, with constant coefficients

$$\begin{cases} y_1' = a_{11}y_1 + a_{12}y_2 + f_1(x), \\ y_2' = a_{21}y_1 + a_{22}y_2 + f_2(x). \end{cases} \quad (3.4.52)$$

If both a_{12}, a_{22} are zero, then the system is already reduced to two first order ODEs. Suppose that $a_{12} \neq 0$. Then from the first equation (3.4.52) we get

$$y_2 = \left[y_1' - a_{11}y_1 - f_1(x) \right] / a_{12}. \quad (3.4.53)$$

We differentiate this once

$$y_2' = \left[y_1'' - a_{11}y_1' - f_1'(x) \right] / a_{12} \quad (3.4.54)$$

and we replace (3.4.53) and (3.4.54) in the second equation (3.4.52), to get

$$y_1'' - a_{11}y_1' - f_1'(x) = a_{12}a_{21}y_1 + \\ + a_{22}\left[y_1' - a_{11}y_1 - f_1(x) \right] + a_{12}f_2(x). \quad (3.4.55)$$

Rearranging terms, we deduce

$$y_1'' + \underbrace{(-a_{11} - a_{22})}_{b_1} y_1' + \underbrace{(a_{11}a_{22} - a_{12}a_{21})}_{b_2} y_1 = \\ = \underbrace{f_1'(x) + a_{12}f_2(x) - a_{22}f_1(x)}_{F_1(x)}, \quad (3.4.56)$$

or else

$$\boxed{y_1'' + b_1 y_1' + b_2 y_1 = F_1(x)}, \quad (3.4.57)$$

which is the ODE we are looking for.

Example

Consider the system (3.3.2)

$$\begin{cases} y' = z + e^{2x}, \\ z' = y. \end{cases} \qquad (3.4.58)$$

Differentiating once the first equation, we obtain

$$y'' = z' + 2e^{2x}; \qquad (3.4.59)$$

from the second equation (3.4.58) it follows that $y'' = y + 2e^{2x}$, or

$$Ly \equiv y'' - y = 2e^{2x}, \qquad (3.4.60)$$

i.e., a linear non-homogeneous second order ODE with constant coefficients. The general solution of the associated homogeneous equation $Ly = 0$ was already deduced in section 2.2; it is

$$y_{\text{homogeneous}} = k_1 e^x + k_2 e^{-x}.$$

As 2 is not a root of the associated characteristic equation, we can search for a particular solution of (3.4.60) of the form

$$Y = A e^{2x}.$$

Replacing this in (3.4.60), we get $A = \dfrac{2}{3}$, therefore $Y = \dfrac{2}{3} e^{2x}$. The general solution of (3.4.60) is thus

$$y = k_1 e^x + k_2 e^{-x} + \frac{2}{3} e^{2x}. \qquad (3.4.61)$$

To get z, we take the first equation of (3.4.58):

$$\begin{aligned} z &= y' - e^{2x} = k_1 e^x - k_2 e^{-x} + \frac{4}{3} e^{2x} - e^{2x} = \\ &= k_1 e^x - k_2 e^{-x} + \frac{1}{3} e^{2x}. \end{aligned} \qquad (3.4.62)$$

Hence the expressions found for y and z are

$$y = k_1 e^x + k_2 e^{-x} + \frac{2}{3} e^{2x},$$
$$z = k_1 e^x - k_2 e^{-x} + \frac{1}{3} e^{2x},$$
(3.4.63)

which coincides with (3.3.9).

EXERCISES AND PROBLEMS

1. Integrate the following systems of homogeneous linear differential systems with constant coefficients, also aplying the conditions (where required):

I. The characteristic equation of the system allows real and distinct roots:

a) $\begin{cases} \dfrac{dy_1}{dx} = y_1 + 4y_2 \\ \dfrac{dy_2}{dx} = y_1 + y_2 \end{cases}$ A: $\begin{cases} y_1(x) = K_1 e^{-x} + K_2 e^{3x} \\ y_2(x) = -\dfrac{1}{2} K_1 e^{-x} + \dfrac{1}{2} K_2 e^{3x} \end{cases}$

b) $\begin{cases} \dfrac{dy_1}{dx} = 3y_1 + 8y_2 \\ \dfrac{dy_2}{dx} = -y_1 - 3y_2 \end{cases}$ A: $\begin{cases} y_1(x) = K_1 e^{-x} + K_2 e^{x} \\ y_2(x) = -\dfrac{1}{2} K_1 e^{-x} - \dfrac{1}{4} K_2 e^{x} \end{cases}$

c) $\begin{cases} \dfrac{dy_1}{dx} = -2y_1 + y_2 \\ \dfrac{dy_2}{dx} = -4y_1 + 3y_2 \end{cases}$ A: $\begin{cases} y_1(x) = K_1 e^{2x} + K_2 e^{-x} \\ y_2(x) = 4K_1 e^{2x} + K_2 e^{-x} \end{cases}$

d) $\begin{cases} \dfrac{dy_1}{dx} = -ay_2 \\ \dfrac{dy_2}{dx} = -ay_1 \end{cases}$ \quad A: $\begin{cases} y_1(x) = K_1 e^{ax} + K_2 e^{-ax} \\ y_2(x) = K_1 e^{-ax} - K_2 e^{ax} \end{cases}$

f) $\begin{cases} \dfrac{dy_1}{dx} = y_1 + 2y_2 - 3y_3 \\ \dfrac{dy_2}{dx} = -2y_1 + y_2 + y_3 \\ \dfrac{dy_3}{dx} = -y_1 + 2y_2 - y_3 \end{cases}$, \quad A: gen. sol.:

$y_1(0) = 2, y_2(0) = 9, y_3(0) = 8$

$\begin{cases} y_1(x) = K_1 + K_2 e^{-x} + K_3 e^{2x} \\ y_2(x) = K_1 + \dfrac{1}{2} K_2 e^{-x} - 7 K_3 e^{2x} \\ y_3(x) = K_1 + K_2 e^{-x} - 5 K_3 e^{2x} \end{cases}$

Sol. of the Cauchy probl.:

$\begin{cases} y_1(x) = 1 + 2e^{-x} - e^{2x} \\ y_1(x) = 1 + e^{-x} + 7e^{2x} \\ y_1(x) = 1 + 2e^{-x} + 5e^{2x} \end{cases}$

e) $\begin{cases} \dfrac{dy_1}{dx} = -y_1 + 8y_2 \\ \dfrac{dy_2}{dx} = y_1 + y_2 \end{cases}$ \quad A: $\begin{cases} y_1(x) = 2K_1 e^{3x} - 4K_2 e^{-3x} \\ y_2(x) = K_1 e^{3x} + K_2 e^{-3x} \end{cases}$

g) $\begin{cases} \dfrac{dy_1}{dx} = -y_1 + y_2 + y_3 \\ \dfrac{dy_2}{dx} = y_1 - y_2 + y_3 \\ \dfrac{dy_3}{dx} = y_1 + y_2 + y_3 \end{cases}$ \quad A: $\begin{cases} y_1(x) = K_1 e^{-x} + K_2 e^{2x} + K_3 e^{-2x} \\ y_2(x) = K_1 e^{-x} + K_2 e^{2x} - K_3 e^{-2x} \\ y_3(x) = -K_1 e^{-x} + 2K_2 e^{2x} \end{cases}$

h) $\begin{cases} \dfrac{dy_1}{dx} = y_1 - y_2 + y_3 \\ \dfrac{dy_2}{dx} = y_1 + y_2 - y_3 \\ \dfrac{dy_3}{dx} = 2y_1 - y_2 \end{cases}$ A: $\begin{cases} y_1(x) = K_1 e^x + K_2 e^{2x} + K_3 e^{-x} \\ y_2(x) = K_1 e^x - 3K_3 e^{-x} \\ y_3(x) = K_1 e^x + K_2 e^{2x} - 5K_3 e^{-x} \end{cases}$

II. The characteristic equation of the system allows complex roots:

a) $\begin{cases} \dfrac{dy_1}{dx} + 7y_1 - y_2 = 0 \\ \dfrac{dy_2}{dx} + 2y_1 + 5y_2 = 0 \end{cases}$ A: $\begin{cases} y_1(x) = e^{-6x}(K_1 \cos t + K_2 \sin t) \\ y_2(x) = e^{-6x}(K_1 + K_2)\cos t + \\ \qquad + e^{-6x}(K_1 - K_2)\sin t \end{cases}$

b) $\begin{cases} \dfrac{dy_1}{dx} = y_1 - 4y_2 \\ \dfrac{dy_2}{dx} = y_1 + y_2 \end{cases}$ A: $\begin{cases} y_1(x) = -2e^x(K_1 \sin 2x - K_2 \cos 2x) \\ y_2(x) = e^x(K_1 \cos 2x + K_2 \sin 2x) \end{cases}$

c) $\begin{cases} \dfrac{dy_1}{dx} = y_1 - 2y_2 \\ \dfrac{dy_2}{dx} = y_1 - y_2 \end{cases}$ A: $\begin{cases} y_1(x) = 2K_1 \cos x + 2K_2 \sin x \\ y_2(x) = (K_1 - K_2)\cos x + \\ \qquad + (K_1 + K_2)\sin x \end{cases}$

d) $\begin{cases} \dfrac{dy_1}{dx} = -9y_2 \\ \dfrac{dy_2}{dx} = 4y_1 \end{cases}$ A: $\begin{cases} y_1(x) = K_1 \cos 6x + K_2 \sin 6x \\ y_2(x) = -\dfrac{2}{3}(K_1 \cos 6x - K_2 \sin 6x) \end{cases}$

e) $\begin{cases} \dfrac{dy_1}{dx} = 12y_1 - 5y_2 \\ \dfrac{dy_2}{dx} = 5y_1 + 12y_2 \end{cases}$ A: $\begin{cases} y_1(x) = e^{12x}(K_1 \cos 5x + K_2 \sin 5x) \\ y_2(x) = e^{12x}(-K_1 \sin 5x + K_2 \cos 5x) \end{cases}$

III. The characteristic equation of the system allows multiple roots:

a) $\begin{cases} \dfrac{dy_1}{dx} = -3y_1 - y_2 \\ \dfrac{dy_2}{dx} = y_1 - y_2 \end{cases}$ A: $\begin{cases} y_1(x) = K_1 e^{-2x} + K_2 x e^{-2x} \\ y_2(x) = -(K_1 + K_2) e^{-2x} - K_2 x e^{-2x} \end{cases}$

b) $\begin{cases} \dfrac{dy_1}{dx} = 3y_1 + y_2 \\ \dfrac{dy_2}{dx} = -4y_1 - y_2 \end{cases}$ A: $\begin{cases} y_1(x) = e^x(K_1 x + K_2) \\ y_2(x) = e^x(K_1 - 2K_2 - 2K_2 x) \end{cases}$

c) $\begin{cases} \dfrac{dy_1}{dx} = (a+1)y_1 - y_2 \\ \dfrac{dy_2}{dx} = y_1 + (a-1)y_2 \end{cases}$ A: $\begin{cases} y_1(x) = e^{ax}(K_1 x + K_2) \\ y_2(x) = e^{ax}(K_1 x + K_2 - K_1) \end{cases}$

2. Integrate the following non-homogeneous ODSs with constant coefficients and find the particular solution in two ways:

A. by using the variation of parameters;

B. by shaping it similarly to the free term.

a) $\begin{cases} \dfrac{dy_1}{dx} + 5y_1 - 2y_2 = e^x \\ \dfrac{dy_2}{dx} - y_1 + 6y_2 = e^{2x} \end{cases}$ A: $\begin{cases} y_1(x) = 2K_1 e^{-4x} - K_2 e^{-7x} + \\ \quad + \dfrac{7}{40} e^x + \dfrac{1}{27} e^{2x} \\ y_2(x) = K_1 e^{-4x} + K_2 e^{-7x} + \\ \quad + \dfrac{1}{40} e^x + \dfrac{7}{54} e^{2x} \end{cases}$

b) $\begin{cases} \dfrac{dy_1}{dx} - y_1 - y_2 = x \\ \dfrac{dy_2}{dx} + 4y_1 + 3y_2 = 2x \end{cases}$ A: $\begin{cases} y_1(x) = (K_1 + K_2 x) e^{-x} + 5x - 9 \\ y_2(x) = (K_2 - 2K_1 - 2K_2 x) e^{-x} - \\ \quad - 6x + 14 \end{cases}$

REFERENCES

1. BÂRZĂ, I., *Analiză Matematică. Culegere de Probleme Rezolvate* (Mathematical Analysis. A collection of solved problems), Niculescu Printing House, Bucharest, 2002.
2. CIORĂNESCU, N., *Curs de Algebră şi Analiză Matematică* (Course of Algebra and Analysis), Ed. Tehnică, Bucharest, 1958.
3. CRAW, I., *Advanced Calculus and Analysis*, Univ. of Aberdeen, 2000.
4. DANKO, P.E., POPOV, A.G., *Vîsşaia Matematika v uprajneniah i zadachah*, Vîsşaia Şkola, Moscva, 1964.
5. COURANT, R., *Differential & Integral Calculus*, t.2, Blackie and Son Ltd, London and Glasgow, 1936.
6. NIŢĂ, A., STĂNĂŞILĂ, T., *1000 de probleme rezolvate şi exerciţii fundamentale pentru studenţi şi elevi* (1000 solved problems and fundamental exercices for students), Ed. BIC ALL, Bucharest, 1997.
7. PĂLTINEANU, G., MATEI, P., *Ecuaţii diferenţiale şi ecuaţii cu derivate parţiale cu aplicaţii* (Ordinary and Partial Differential Equatins with aplications), Matrix Rom, Bucharest, 2007.

8. SOARE, M.V., TEODORESCU, P.P., TOMA, I., *Ordinary differential equations with applications to mechanics*, Springer, Dordrecht, 2006.
9. STRANG, G., *Calculus*, Wellesley College, 1991.
10. TOMA, I., *Analyse Mathématique. Calcul différentiel*, (Mathematical Analysis. Differential Calculus) Conspress, Bucharest, 2010.
11. TOMA, I., MOŞNEGUŢU, V., *Analiză Matematică. Ecuaţii diferenţiale ordinare. Calcul integral* (Mathematical Analysis. Ordinary differential equations. Integral Calculus), Ed. Conspress, Bucharest, 2011.
12. Links:

http://www.springer.com/math/analysis/book/978-3-540-88704-1

http://ocw.mit.edu/OcwWeb/hs/home/teachers/index.htm

http://en.wikibooks.org/wiki/Calculus

http://www.springer.com/mathematics/analysis/journal/526

http://www.springer.com/mathematics/journal/11957

www.ingramcontent.com/pod-product-compliance
Lightning Source LLC
Chambersburg PA
CBHW071426180526
45170CB00001B/232